★ ★ ★ ★ ★ ★ ★ ★ art of the ★ ★ ★ ★ ★ ★ ★ ★ ★

CHICKEN COOP

★ ★

art of the CHICKEN COOP

A Fun and Essential Guide to Housing Your Peeps

Chris Gleason

Fox Chapel
PUBLISHING

Acknowledgments

A lot of people were involved in helping this book to become a reality. I would like to start by thanking Wasatch Community Gardens for promoting the cause of backyard chickens in Northern Utah. I would also like to thank Kerri Landis for doing such a smart and careful job as an editor, and Peg Couch at Fox Chapel Publishing earned my appreciation for taking a chance with a new author. Here's to more great projects down the road!

ISBN 978-1-56523-542-7

Library of Congress Cataloging-in-Publication Data

Gleason, Chris, 1973-
 Art of the chicken coop / Chris Gleason.
   p. cm.
Includes index.
ISBN 978-1-56523-542-7
1. Chickens--Housing. I. Title.
SF494.5.G54 2011
636.5'0831--dc22

                    2010051994

To learn more about the other great books from Fox Chapel Publishing, or to find a retailer near you, call toll-free 800-457-9112 or visit us at *www.FoxChapelPublishing.com*.

Note to Authors: We are always looking for talented authors to write new books in our area of woodworking, design, and related crafts. Please send a brief letter describing your idea to Acquisition Editor, 1970 Broad Street, East Petersburg, PA 17520.

Printed in China
First printing: June 2011

From the Author

Having grown up on a dairy farm in upstate New York, it wasn't a large stretch for me to begin raising chickens as an adult. My wife and I caught the bug about six years ago, and have been backyard poultry enthusiasts and evangelists ever since. We have had anywhere from six to twenty-five birds at a time. In addition to building and selling coops, I do a lot of consulting with small-flock owners in our area. This has led me to teach chicken coop workshops in conjunction with Wasatch Community Gardens. More than two hundred people attended my last class, and it has become an annual event. I very much enjoy the daily pleasures of raising chickens and helping others to get involved. Welcome to one of the most interesting and rewarding backyard hobbies you can find!

⬆ Here's one of my coops, chock-full of our current flock. Though we don't name our chickens, we really enjoy watching them go about their daily business.

➡ My daughter, Abigail, has grown up helping us raise chickens—one of the perks of this hobby is being able to show your kids firsthand where some of their food comes from.

Chris Gleason is the author of several books for the DIY market including *Built-In Furniture for the Home, Kitchen Makeovers for Any Budget, Complete Custom Closet, Old-School Woodshop Accessories,* and *Building Real Furniture for Everyday Life.* He currently builds and sells chicken coops, and has owned Gleason Woodworking Studio for more than 13 years.

CONTENTS

ABOUT THIS BOOK

This book is primarily a handbook to guide your creation of the perfect chicken coop for your backyard flock. There are plans and step-by-step illustrated instructions for constructing seven unique coops, and a brief gallery for more inspiration.

the coops

Just over half of the coop designs are sized to fit a flock of 6—I figured this was a good size that most people would use. There are also designed options for 12, 14, and 15 chickens. However, if you see a coop you like, but it isn't sized correctly for your flock, the easiest method for resizing is simply enlarging the floor area of the coop by 2 square feet (.2 square meters) per additional bird and then adapting the other parts to fit. You will also need 8" to 10" (200mm to 250mm) of roost space per bird; and don't forget that you need one nest box for every 4 chickens.

I like to use as much salvaged and recycled construction material as possible, so you'll often see me utilizing old cabinet doors as coop doors, partial sheets of plywood rescued from another project, or even old hardware and fixtures. I encourage you to piece together your coop using creativity and salvaged materials, but of course you can always go to the nearest lumber supply store and pick up new plywood and 2x4s.

After you've finished building your coop of choice, turn to page 136 to read up on building a run. You'll need one of those unless you already have a fenced-in area for your flock, or if you're comfortable allowing your chickens to really be free range!

HOW MANY CHICKENS FIT IN THESE COOPS?

- 6 chickens: Coop #2, #3, #4, #6
- 12 chickens: Coop #1
- 14 chickens: Coop #5
- 15 chickens: Coop #7

A Word About Metric

The measurements in this book appear first in imperial (inches, feet), followed by metric (millimeters), in order to accommodate woodworkers who use both measurement systems. When the measurement must be exact—as when one piece must fit with another precisely—we've rounded to the nearest whole millimeter. When the measurement is more general, we've rounded to the nearest 0, 2, or 5mm. With regard to nominal lumber measurements (2x4, 1x6, etc.), we've listed the metric actual measurement—in other words, the measurement that the piece of wood actually is, rather than what it was originally sawn to. By this reasoning, a 2x4 is 38x89mm, rather than 50x100mm. Additionally, plywood thickness is shown as the most common actual metric thickness; for example, though ½"-thick plywood can be anywhere from 11 to 12½mm, the most common thickness (and how we will list it) is 11mm.

advice and entertainment

Building a coop is not all instructions and assembly—you need inspiration and encouragement. And let's not forget why you're embarking on this endeavor—you want to raise chickens! For these reasons, I tossed in a lot of fun and useful information to help you on your way toward being a good flock parent. There are photo- and advice-filled profiles from three chicken keepers who've lived the chicken-keeping life. Also, keep your eyes peeled for the sidebars and tidbits on the bottom of each right-hand page throughout the text. This informational egg hunt will yield interesting, helpful, and just-plain-entertaining information about raising chickens, egg dishes from around the world, advice on selecting breeds, and more!

coop-a-doodle-do:

COOP DESIGNS

Which came first: the **chicken or the egg?** The correct answer to this age-old question is neither. **The coop came first.**

This section contains seven coop designs to get you on your way to housing your peeps. Whether you have a large amount of space or a small backyard, six birds to house or fifteen, lots of money to spend or little—you'll find a design here to help your chickens get cooped up in style!

When you've finished your coop, flip to page 136 for information on building a run to fence in your chickens.

Important Coop Parts

- ❏ Roosts: To sit on
- ❏ Ladders: To climb up
- ❏ Nest Boxes: To lay eggs in
- ❏ Floor: To stand and poop on
- ❏ Windows and Doors: To ventilate and allow easy cleaning
- ❏ Roof: To keep the weather out
- ❏ Run: To keep predators out and keep chickens from wandering away

ESSENTIAL DIMENSIONS FOR COOP BUILDING

When resizing a coop design to fit your flock, pay special attention to the following dimensions:
- 2 square feet (.2 square meters) of coop space per bird
- 1 nest box per 4 birds
- 8" to 10" (200mm to 250mm) of roost space per bird

SUNNY SIDE UP
A Classic Design with an Easily Accessible Nest Box

I've built a number of coops in this style, and they've worked out really well. I like to dress them up a bit, but I'm sure different folks will enjoy personalizing their coops to match their own unique tastes. Regardless of the aesthetic treatments you might use, there are a number of practical qualities to this design that any poultry owner will appreciate.

In terms of size alone, this design is pretty versatile. The coop measures 4' x 6' (1200 x 1800mm), which provides 24 square feet (2.2 square meters) of floor space. With an allotment of 2 square feet (.2 square meters)/bird, a structure of this size could house up to 12 chickens. You could certainly scale the concept to accommodate smaller or larger flocks, too.

The nest boxes are contained in a separate area—often called a sidecar—that is easily accessible from outside of the run. Some people like this aspect, while others don't seem to mind entering the run every day to collect eggs. I fall into this latter camp, but then again, I am somebody who likes to pull a folding chair into the run daily and relax with the chickens.

Because this coop is up on stilts, its large footprint provides a big shady area for birds to escape to on hot days. This also makes for good airflow; the area below the coop receives a breeze to keep the ground dry, which is an important consideration for disease prevention. And speaking of ventilation, the coop has two large sets of double doors that allow for great interior circulation. The doors also make clean-out easier, and the roosts are removable to help with this as well.

Because I often build my coops off-site, I need to be able to transport them to their final homes. In this case, the coop was a skosh too large for my van, so I built it to be easy to take apart and re-assemble. This may not be a consideration for you, but it is a real-world fact of life for some of us.

Shakshuka, Israeli: Eggs poached in spicy tomato sauce.

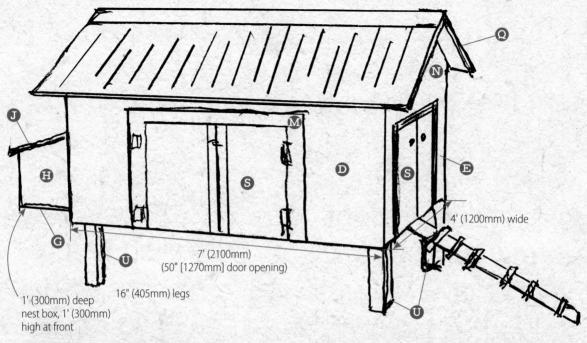

4' (1200mm) wide

7' (2100mm)
(50" [1270mm] door opening)

16" (405mm) legs

1' (300mm) deep
nest box, 1' (300mm)
high at front

1 **Start with the bottom.** The place to start building a coop like this is with the bottom (A). I used a 4' x 8' (1200 x 2400mm) sheet of ⅝" (15mm) plywood, although any thickness between ½" (11mm) and ¼" (6mm) would be fine. The bottom of the coop will be 4' x 6' (1200 x 1800mm), so I needed to trim down the plywood—I used a straightedge to draw a line for the cut.

2 **Cut out the bottom.** I used a jigsaw, since it tends to be my go-to saw, but a circular saw would be a great choice too. Keeping the plywood up on sawhorses makes the process a bit easier.

MATERIALS LIST

item	material	dimensions	quantity
Ⓐ Platform (floor)	⅝" (15mm) plywood	4' x 6' (1200 x 1800mm)	1
Ⓑ Long-side floor framing	2x4 (38 x 89mm)	6' (1800mm)	2
Ⓒ Short-side and middle floor framing	2x4 (38 x 89mm)	45" (1100mm)	3
Ⓓ Side wall panels	½" (11mm) plywood	6' x 3' (1800 x 900mm)	2*
Ⓔ Front and back wall panels	⅝" (15mm) plywood	4' x 3' (1200 x 900mm)	2*
Ⓕ Wall corner posts	2x2 (38 x 38mm)	32" (800mm)	4
Ⓖ Nest box bottom	¾" (17mm) plywood	43½" x 12" (1100 x 300mm)	1
Ⓗ Nest box sides	¾" (17mm) plywood	1' x 1' (300 x 300mm)	2
Ⓘ Nest box back	¼" (6mm) plywood	45" x 10" (1100 x 300mm)	1
Ⓙ Nest box lid	¾" (17mm) plywood	46" x 12" (1200 x 300mm)	1
Ⓚ Long nest box divider	¾" (17mm) plywood	12" x 10" (300 x 300mm)	1
Ⓛ Short nest box dividers	¾" (17mm) plywood	6" x 10" (150 x 250mm)	2
Ⓜ Side door trim	1x4 (19 x 89mm) pine	various lengths	10 linear feet
Ⓝ Gable end panels	¾" (17mm) OSB	48" x 14" (1200 x 400mm)	2
Ⓞ Truss	2x2 (38 x 38mm)	48" (1200mm)	2
Ⓟ Truss cross-tie	¼" (6mm) plywood	48" x 10" (1200 x 300mm)	1
Ⓠ Roof panels	⅜" (8mm) plywood	84" x 40" (2100 x 1000mm)	2
Ⓡ Gable trim	2x2 (38 x 38mm)	40" (1000mm)	2
Ⓢ Door panels	¾" (17mm) MDF or similar	28" x 16" (700 x 400mm)	4
Ⓣ Sunburst overlay	¼" (6mm) plywood	48" x 14" (1200 x 400mm)	1
Ⓤ Legs	2x4 (38 x 89mm)	18" (500mm)	4
Scrap metal			
Screws			
Hinges			
Paint			
Brass hook and catch			
Tar paper			
Galvanized roofing			
Chicken wire			
Staples			

* Note that I didn't have pieces of these dimensions for this coop, so I patched together smaller pieces and reinforced the joints as needed. Follow whatever approach works for you.

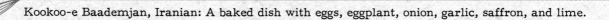

Kookoo-e Baademjan, Iranian: A baked dish with eggs, eggplant, onion, garlic, saffron, and lime.

3 Screw on the 2x4 frame. The stiffness of the floor comes mostly from the framing (B, C) along its edges. 2x4s provide a great deal of rigidity when they're screwed to the plywood with 2" (50mm) screws at 12" (300mm) intervals. Attach the 3 shorter 2x4s (C) on the two short edges and the middle of the floor (A).

4 Screw the 2x4 corners. To get a tight joint on the corners, I used a long pipe clamp and screwed the 2x4 corners together. Pre-drilling wasn't necessary. You'll notice that the wood is used—I love to build coops from reclaimed materials, and the patina on this wood doesn't affect its strength or structural integrity.

5 Make the solid side panel. I made one solid side (D) from ½" (11mm) plywood with 2x2s screwed to the underside at the vertical edges. The 2x2s (F) will act as corner posts to attach the adjacent side panels to. I set up the 2x2s 4" (100mm) from the bottom of the plywood so that the side panels will hang down and cover up the 2x4 of the bottom assembly.

6 Attach the solid side panel. One side panel (D) can be screwed to the bottom assembly with 1½" (40mm) screws. This joint can be reinforced with construction adhesive if you'd like.

3

4

5

6

7 Make the front panel. The front of the coop (E) was made from a pair of ½" (11mm) plywood panels. I used two panels because that was what I had to work with. In an ideal world, I would've used a single piece, but with reclaimed materials, sometimes you have to make compromises. This photo shows the first panel set in place.

8 Prepare to cut in door openings. The front of the coop (E) requires a large set of doors (S) to provide easy access for cleaning and good ventilation. I laid out the opening with a framing square. Sometimes I build chicken coop doors out of old cabinet doors, which require a precisely-sized opening. In this case, I'll be building the doors later, so the exact size of the opening isn't critical.

9 Cut front door openings. I used a jigsaw to cut out the opening on the first panel and then drew lines for a matching cut on the other panel.

10 Reinforce the front panels. When the opening is cut out, it is important to reinforce the joint between the two panels. On the interior, I screwed a piece of scrap with plenty of surface area for screws and glue. This piece was 20" (510mm) long, leaving 10" (255mm) on either side of the joint to work with.

The first documented use of the word "squawk" was in 1821.

11

12

11 Construct the first back panel. The back of the coop (E) was assembled in a similar fashion. A major difference was the size of the opening—the nest box will be installed here. Again, I could've used a single plywood panel, and then made a cutout for the entrance into the nest boxes, but since I didn't have large pieces of plywood to work with, I began by setting a vertical piece into place to build up the back.

12 Construct the second back panel. The other vertical panel that makes up the back (E) was constructed just like the solid side panel (D), using a 2x2 corner post (F).

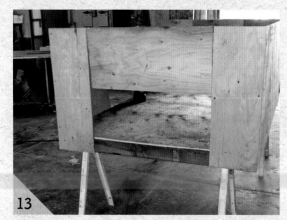

13

13 Construct the last back panel. With both vertical panels in place, I measured for the middle horizontal panel and cut it out on the table saw.

14 Reinforce the back panel. The middle panel has to be reinforced on the inside just like the joint on the front panel. This construction method provides plenty of strength, while still allowing you to be resourceful with the materials you have on hand.

14

15

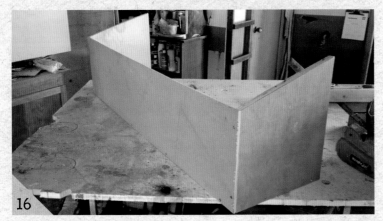

16

17

18

15 Construct the side door panel. The other side of the coop (D) went together just like the back panel. Again, I could've used a large plywood panel for this, but—you guessed it—I didn't have any on hand, and this method allowed me to use up odds and ends that otherwise might've ended up in a landfill. This side is perfect for scraps because it also has a large set of doors (S) to aid with cleanup.

16 Construct the nest box. The external nest box is essentially a four-sided box that is constructed with butt joints. The sides (H) are cut so that they taper toward the outside of the box, and the whole subassembly is open on the back side so that the chickens will have easy access to the nest boxes from the coop interior. I used ¾" (17mm)-thick plywood scraps for the nest box sides (H) and bottom (G).

17 Attach the nest box. I had some scrap sheet metal that I used to fabricate the simple L-shaped brackets that attach the nest box to the coop. Tin snips, a metal-cutting jigsaw blade, or an angle grinder will help you to cut the metal to the right size, or you could just use pre-made brackets.

18 Notch the bottom. The nest box is also supported by the floor of the coop—I cut out the edges of the bottom of the nest box so that the floor could be set into the opening in the side panel. This makes for a really strong connection between the nest box and the main structure.

Provide 1 nest box for every 4 birds.

19 Install nest box dividers. Because this nest box measures nearly 4' (1200mm) long, I could've divided it into 4 separate compartments (12" [300mm] wide is considered sufficient for most layers), but I divided it into 2 main parts using a long divider (K) and then partially divided the halves with shorter dividers (L). I like to build nest boxes on the big side—in my experience, it isn't uncommon for multiple hens to occupy a box at the same time, so I figured a little extra space wouldn't hurt.

20 Attach trim to the front. Once the main structural elements are in place, I turn my attention to the details. I like to make and attach fancy trim to the most visible sides of the coop. In this case, the horizontal trim piece (M) above the side door opening also serves to reinforce the joints between the plywood sub-panels. You can also see the ladder here.

21 Cut out the first gable. The roof of this coop is pretty straightforward—it has identical gables (N) on the front and back. I constructed it by cutting out gable profiles from ¾" (19mm) oriented strand board (OSB), and I decided on the exact angle (pitch) of the roof by just eyeballing it rather than doing any math. If you want to get out an angle finder and approach it that way, go for it.

22 Notch the bottom edge of the gable. The bottom edge of the gable panel (N) needed to be notched with a jigsaw so that it would fit over the multiple pieces making up the front panel (E).

23 Cut out the second gable. The first gable panel can be used as a template for the second one. Be careful, however, because the notches on the bottom may not be the same—make sure to trace the cut-outs you need so the gable fits correctly.

19

20

EGG CUSTARD

2 cups whole milk	⅓ cups sugar
2 eggs	1 tsp vanilla extract
2 egg yolks	Dash of nutmeg

Simmer milk over medium-low heat. Combine eggs, yolks, sugar, and vanilla; pour this mixture gradually into the milk. Strain the mixture into 6 ramekins; top with nutmeg. Bake in a water bath in a 300° F (150° C) oven for about 30 minutes. Cool 1 hour before serving.

21

22

23

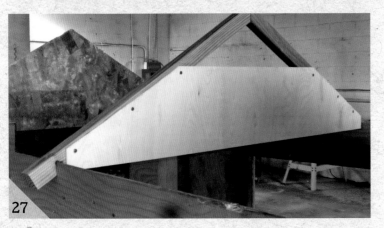

24 Screw on the gables. The notches don't have to fit snugly—a bit of wiggle room is fine. After I had the gable panels (N) in place, I screwed them into the front and back panels (E).

25 Begin building the truss. The gable panels do a great job supporting the roof on its ends, but some additional structure will be required in the middle. I built a truss (O) from 2x2s. I started out by clamping the 2x2s to a gable's edge. This made it easy to verify that I had a good match.

26 Add a plywood cross-tie to the truss. The strength of the truss comes from the rigid cross-tie (P) made from ¼" (6mm) plywood. To make the cross-tie, I set the plywood into place and traced the profile of the truss (O) onto it. Cutting along the lines made the cross-tie fit perfectly without having to measure. Screw the cross-tie in place.

27 Notch the truss 2x2s. The truss has to be notched at the bottom so that it can slip onto the top edge of the walls. This is easy to do—just hold the truss in place above the coop and mark the place where the walls intersect it.

Quiche Lorraine, French: Eggs, cream, and bacon baked in a pastry crust.

28

29

28 Hinge the roof panels. The roof (Q) is made of ⅜" (8mm) plywood. To make it easy to handle on site, I joined the two halves with a pair of inexpensive hinges.

29 Dry-fit the roof. The hinges aren't necessary, but having them makes lining up the roof as a unit quick and easy. Here, you can see the overhang at the front and back of the coop, which needs to be reinforced for both structural and aesthetic reasons.

30 Add the gable trim. I used a pair of 2x2s to build out the edge of the roof. This gable trim (R) stiffens up the roof (Q) and makes it look more substantial.

31 Paint the coop. A coat of paint brightens up the reclaimed lumber, which looked a little grim beforehand.

30

31

32

32 Add the nest box lid. Cover the nest boxes with a simple flap (J) made from ¾" (17mm) plywood. You can raise and lower it easily to collect eggs and clean out the nests. Attach it with a pair of standard hinges.

33 Attach the front doors. I made the front doors (S) from some scrap MDF, which will hold up fine outdoors if it is well-finished. The pieces that I had weren't big enough to span the whole opening, which seems like a problem, but I had a solution in mind for this.

34 Bridge the door gap. I bridged the gap with a simple piece of ½" (11mm) plywood that I screwed to one of the doors. It overlaps onto the other door and creates a nice, tight seal. I added a brass hook catch to keep the doors shut.

33

BLUE EGGS?

Araucana and Ameraucana chickens lay blue eggs, seen here in the middle, as compared to typical brown and white colors.

34

Ameraucanas and Aracaunas lay light blue or light green eggs.

35

36

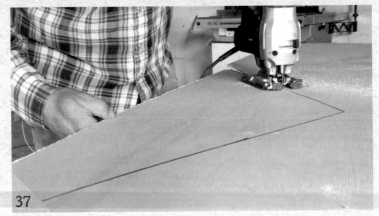

37

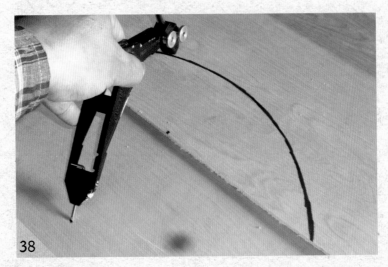

38

35 Decorate the door. Rather than apologize for this solution, I use it as an opportunity for some adornment: a hand-printed chicken emblem seemed right at home here.

36 Trace gable cover panels. The OSB used for the gable panels is structurally sufficient, but it doesn't look amazing, so I wanted to dress up the front a bit. To cover it up with a properly fitting panel (T), I just traced the profile on to a ¼" (6mm) plywood blank. In the next few steps, we'll transform this panel into a sunrise overlay.

37 Cut out the gable cover panel. I used a jigsaw to cut out the panel (T), but a circular saw would've worked fine too.

38 Lay out the sunrise motif. I often dress up the front gable of my coops with a sunrise overlay. To create this motif, I began by laying out an arc—the sun—at the bottom of the gable cover panels (T). In my experience, it sometimes fits best to set the center of the circle below the edge of the panel so that the sun is not a half-circle, but rather an arc that is shorter than it is wide.

39

40

41

39 Lay out the rays. Lay out the rays of the sun in whatever arrangement looks best. I start with the center ray, and then space them out so they look right. Using pencil isn't a bad idea at first, because it is easy to change your mind and redo the spacing.

40 Attach the sunrise overlay. Cut away the waste material between the rays of the sun. Glue and screw the overlay (T) to the OSB gable panel (N). Some contrasting paint really brightens up the gable, too.

41 Disassemble the coop. If you're transporting the coop somewhere, disassemble it and pack the parts into a van or truck. Reassembly takes less than an hour—and it's much easier than renting a truck and getting 4 people together to haul the finished coop around.

Another word for egg white is albumen.

42

42 Attach the legs. To attach the legs, I suggest turning the bottom assembly upside down. You can then run beefy screws through the legs (U) into the frame. Then, flip this subassembly right-side up and attach the wall panels (D,E).

43 Attach the ladder. The chicken ladder is a cute addition to any coop. Younger birds in particular seem to use it quite often. I attached the ladder to the coop by screwing on a small hinge from below.

PAD THE NEST BOX

You risk broken eggs if you don't cushion your nest boxes with some kind of soft bedding material, such as straw, woodchips, shredded newspaper, or sawdust.

43

44

44 Put on the roof. As I mentioned before, having the roof (Q) hinged together helps to streamline the on-site experience. Insert screws through the roof (Q) into the gables (N) and truss (O).

45 Add the roof top cap. To keep the birds dry, I covered the roof with a layer of tar paper and galvanized roofing. The top cap is an essential part of the system. Use staples to attach the tar paper, and exterior-grade screws for the tin.

45

Çilbir, Turkish: Poached eggs over garlic-yogurt sauce, drizzled with butter and paprika.

CHICKEN CONDO
A Multilevel Coop with a Small Footprint

I build a lot of coops with traditional gabled roofs, so once in a while it is fun to play around with different styles. This coop has a simple sloped roof that is easy to build, and I liked the look of the structure as a whole—the combination of the weathered siding and the white trim reminded me of a lot of buildings I saw years ago on a visit to Cape Cod.

The overall functionality of this coop is rather unique, too: the interior is divided into a couple of levels, which provides more usable square footage, and the coop has a door in the floor that allows the residents to come and go as they please. Fencing in the space below the coop and attaching a small wire enclosure on one side creates a very practical setup with a small footprint. The photos and instructions that I've included for the wire enclosure (see page 141) would also make for a pretty handy portable chicken tractor—a bottomless screened-in cage that you move each day to a new plot, allowing the chickens to have a fresh feeding ground and to fertilize a new patch of ground each day.

🐔 Popular dual-purpose breeds include Plymouth Rocks, Wyandottes, and New Hampshires.

1

2

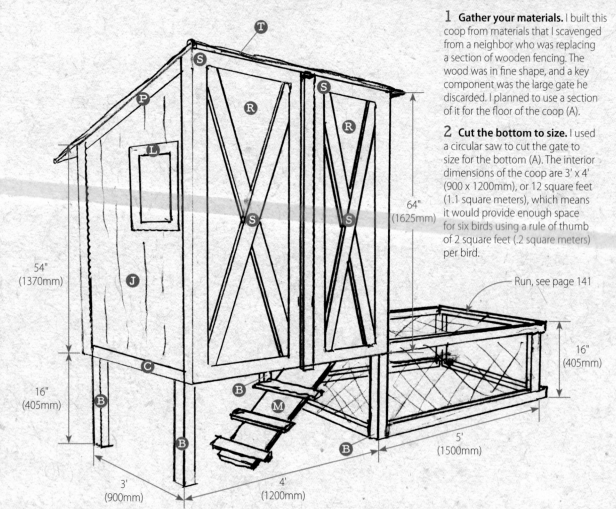

T

S

P

S

R

S

R

L

S

S

J

64"
(1625mm)

54"
(1370mm)

C

B

B

M

B

B

16"
(405mm)

16"
(405mm)

Run, see page 141

5'
(1500mm)

3'
(900mm)

4'
(1200mm)

1 **Gather your materials.** I built this coop from materials that I scavenged from a neighbor who was replacing a section of wooden fencing. The wood was in fine shape, and a key component was the large gate he discarded. I planned to use a section of it for the floor of the coop (A).

2 **Cut the bottom to size.** I used a circular saw to cut the gate to size for the bottom (A). The interior dimensions of the coop are 3' x 4' (900 x 1200mm), or 12 square feet (1.1 square meters), which means it would provide enough space for six birds using a rule of thumb of 2 square feet (.2 square meters) per bird.

MATERIALS LIST

item	material	dimensions	quantity
Ⓐ Bottom panel	¾" (17mm) plywood or lumber	4' x 3' (1200 x 900mm)	1
Ⓑ Long side framing	2x2 (38 x 38mm)	5' (1500mm)	4
Ⓒ Short side framing	2x2 (38 x 38mm)	34½" (876mm)	4
Ⓓ Joining plates	¾" (17mm) plywood	6" x 6" (152mm x 152mm)	8
Ⓔ Back panel	¼" (6mm) plywood	3' x 3' (900 x 900mm)	1*
Ⓕ Interior back cleat	⅝" (16mm) lumber	37" x 5½" (940 x 140mm)	1
Ⓖ Interior right side cleat	⅝" (16mm) lumber	35" x 5½" (890 x 140mm)	1
Ⓗ Interior platform	½" (11mm) plywood	30" x 30" (762 x 762mm)	1
Ⓘ Second floor support	2x2 (38 x 38mm)	20" (510mm)	1
Ⓙ Siding	⅝" (16mm) fence pickets	4' x 6" (1200 x 150mm)	14
Ⓚ Interior window support trim	1x2 (19 x 38mm) pine	5' (1500mm) total	
Ⓛ Window frame	1x3 (19 x 63mm) pine	30' (9100mm) total	
Ⓜ Chicken ladder	1x6 (19 x 140mm) pine	32" (815mm)	2
Steps for ladder	¾" (19mm) scrap	8" x 1¼" (205 x 30mm)	As many as are needed
Ⓝ Nest box, top, bottom, and back	¾" (17mm) plywood	12" x 12" (305 x 305mm)	3 for each nest box
Ⓞ Nest box, sides	¾" (17mm) plywood	13½" x 12" (345 x 305mm)	2 for each nest box
Ⓟ Roof cleats	¾" (19mm) pine	4' x 4" (1200 x 102mm)	2
Ⓠ Header	1x6 (19 x 140mm) pine	4' (1200mm) long	1
Ⓡ Door panel	⅜" (8mm) plywood or similar	64" x 17¾" (1625 x 450mm)	2
Ⓢ Door trim	1x6 (19 x 89mm) pine	34' (10400mm) total	
Ⓣ Roof	Metal panel	54" x 42" (1372 x 1068mm)	1
Ⓤ Hinges	Standard butt hinges	4" (102mm)	4
Ⓥ Roosts	1" (25mm) dowel	4' (1200mm)-long	2
Ⓦ Roost blocks	⅞" (22mm) pine	8" x 5" (205 x 128mm)	2
Exterior-grade screws			
Staples			

*Note that I didn't have pieces of these dimensions for this coop, so I patched together smaller pieces and reinforced the joints as needed. Follow whatever approach works for you.

 The largest recorded chicken egg weighed 12 oz.

3 Prepare 2x2s for framing. I could've used 2x4s for the framing (B, C), but that would've been overkill. Ripping the 2x4s down the middle on my table saw allowed me to very easily double the amount of wood that I had to work with.

4 Cut the subassembly parts. This coop is structured around a matching pair of side subassemblies. Using my homemade 2x2s, I cut two 5' (1500mm)-long vertical pieces (B) and two 34½" (875mm)-long horizontal pieces (C) for each subassembly. I clamped up the sides so that their parts could be fastened together. Place the lower horizontal 2x2 (C) 18" (458mm) up from the bottom.

5 Prepare the joining plates. I could've just screwed the 2x2s together, but this would not have created a particularly strong or durable joint. I like to screw shopmade joining plates (D) onto the joints where the members meet up. And, even though they wouldn't be all that prominent in the final result, I decided to dress them up just a bit. Use whatever design you like.

6 Cut out the joining plates. I cut the braces (D) out on my band saw, although a jigsaw would work fine too. If you prefer a simpler look, you could just lop off one corner of each piece on a mitersaw or a table saw equipped with a miter gage.

3

4

5

6

7 Attach the braces. I made sure to use no fewer than 4 screws on each brace (D). You can reinforce the joint further by running a bead of exterior-grade construction adhesive on the backsides of the braces.

8 Complete the side subassemblies. The completed side subassemblies are lightweight and very strong.

9 Attach the floor. When both sides have been assembled, screw down the floor panel (A) onto the lower horizontal subassembly frame pieces.

10 Examine the coop. At this point, I could see the coop start to look like something. The structure had a tendency to rack (wobble side-to-side), but I wasn't worried—experience has shown me that this weakness will be eliminated with the addition of a solid back panel (E).

11 Prepare the back panel. I like to use ¼" (6mm) plywood for the back of coops because it is extremely rigid but it doesn't add too much weight. If you don't have a sheet wide enough to span from the top to the bottom (I didn't), it is fine to make up the difference with a second piece. Screw on the back panel piece(s) (E).

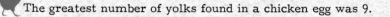

The greatest number of yolks found in a chicken egg was 9.

12 **Attach the interior back cleat.**
To seal up the gap between the two back panels (E), I patched in a cleat (F) on the interior of the coop. I attached it at 20" (510mm), a height that allows it to do double duty and serve as a support for the second floor of the coop.

13 **Attach the interior side cleat.** I installed a similar cleat (G) perpendicular to the back cleat (F) so that I would have another support for the second floor.

14 **Attach the interior platform.**
Here's the second floor (H) in place—there isn't much to it, but it is a simple way of adding a lot of extra usable space to the interior. I used a scrap length of 2x2 (I) to support the cantilevered corner. Screw the platform (H) onto the cleats (F, G) and the vertical support (I).

15 **Attach the lower edge of the siding.** The weathered cedar fence pickets were quite a score, and I was excited about the character they would provide when used as exterior siding (J). To make the installation process easier, I tipped the coop on its side—this meant I didn't have to fight with gravity. My other trick for installing siding involves lining up the boards carefully with the bottom edge of the lower 2x2 and then letting them "run wild" over the top edge. Even though they were different lengths, that made no difference—I knew that I would cut them off after they were screwed down. This method is much faster and easier than cutting each board individually, and it usually provides a neater result, too.

12

13

14

15

16

17

KEEP WATER CLEAN LONGER

Chickens spend a good part of their day scratching around in the dirt on the floor of their run, and the only time this is a problem is when they manage to fill up the basin in their waterer and you have to go clean it out prematurely. To prevent this, most people elevate the waterers a little bit.

16 **Scribe the top of the sides.** When the boards (J) were screwed down, I used a straightedge to scribe a line at the top of the wall—45" (1145mm) from the bottom of the boards in the back, 54" (1370mm) in the front.

17 **Remove the waste material.** Using a circular saw, it took less than 30 seconds to remove the waste material.

18 **Examine the side.** The finished side has the look of an old barn—just what I was going for.

18

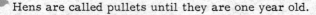

Hens are called pullets until they are one year old.

19 Add a window. After completing the second side, I decided to cut a window opening as a way of bringing some light and fresh air inside. I used a jigsaw and had it cut out in under a minute. If you make the window two boards wide, you only need to cut the top and bottom.

20 Install interior support trim. Because I used pickets for the siding, cutting the window opening now meant that a couple of the boards weren't supported on both ends— I solved this problem by installing a trimmer (K) below the window, and the wall was once again as good as new.

21 Screen in the window. The screen is very easy to attach— I stapled it directly to the siding with ⅜" (10mm) staples.

19

20

21

22

23

24

25

22 **Glue up the window frame.** The window frame (L) is a simple mitered assembly. Use a miter box to cut the pieces at 45° so that the inner frame fits the window size. Clamping miters while the glue dries is a snap using blue painter's tape.

23 **Double-check the frame size.** To make sure the frame (L) is square, I measured the diagonals and made sure the measurement was the same. Any discrepancies can be eliminated with a nudge as required. Paint the frame white when it is dry.

24 **Attach the frame to the coop.** I secured the window frame (L) to the structure with construction adhesive and nails. The white trim around the edges of the side is a good look, too.

25 **Make a chicken ladder.** The second floor of this coop is way too far up for hens to just hop up, so a chicken ladder (M) is essential. A chicken ladder is basically a long board with many short rungs. I like to run screws through the front and back sides of the rungs so they don't loosen up over time.

 Croque-madame, French: Grilled ham and cheese sandwich topped with fried egg.

26

27

26 Construct the nest boxes. Nest boxes (N, O) can be pretty simple—this freestanding, five-sided box can be placed anywhere in the coop. Use screws (be sure to pre-drill) or nails and glue to fasten the 5 pieces together. Make as many nest boxes as you want.

27 Add the header and roof cleats. The metal roofing (T) I used for this coop is very rigid and doesn't need a lot of support beneath it, even in a particularly snowy area—but it does need to be fastened to the coop very tightly. Attach the header (Q). A pair of cleats (P) at the top of the walls provides a good place for attachment.

28 Build the door frame and panel. My approach to building coop doors is influenced by the materials I have on hand at the time. In this case, I used 1x4s (S) to make a rectangular frame that I "skinned" with leftover laminate flooring on the backside. While this is hardly a conventional technique, it is certainly a reasonable solution. Another approach is to prepare your door panels (R) and trim them with the 1x4s (S).

28

29

29 Mark the first slat of the X.

I wanted to embellish the fronts of the doors with a traditional X-shaped pattern (S). Here's my trick for making the slats fit neatly into place with no head-scratching required: simply lay the first slat into position—just so that it looks nice—and then use a straightedge to extend the lines of the door's frame onto the slat.

30 Cut the first slat of the X.

Once you've cut away the waste with a jigsaw or circular saw, the slat should drop neatly into place.

31 Cut the second slat of the X.

To fit the second slat, I mark its corners just like the first slat, and then I use a straightedge to mark its intersection with the first slat. You can see that I identified the waste area with an X—once this section is cut away, the top and bottom halves of the slat should fit just fine, and they can be glued and screwed down.

30

SCOTCH EGGS

Traditional Scottish breakfast and picnic food

6 eggs, hardboiled	1lb bulk sausage
1 cup bread crumbs	½ cup flour
	1 egg, beaten
	Oil for frying

Remove the eggshells. Roll each egg in flour. Press sausage around each egg. Roll the eggs in the beaten egg, and then in the bread crumbs. Refrigerate or freeze the eggs for a bit, and then fry them in the oil until the sausage is browned.

31

 "Bantam" refers to smaller breeds of fowl.

32

33

34

32 Hang the doors. Hanging the completed doors is fairly straightforward—my tip is to fasten the hinges (U) to the door first. I then fasten the top hinge to the coop so that I don't have to fight with gravity (just try screwing in the bottom one first if this doesn't make sense—you'll get it right away). When the top is secured with a screw, I move the bottom hinge from side to side until the door is properly aligned. I then put a screw into the bottom hinge. When both doors are hung and lined up nicely, I fill in the rest of the holes.

33 Attach the roof. The ridges on this metal roof (T) give it quite a bit of stiffness. It is also coated to stand up to the elements. Attach the roof to the roof cleats using exterior-grade screws.

34 Review your work. Here's a shot of the exterior so far.

35

35 Install the run. I chose to build a small run attached to the coop. See page 141 for details. The run can simply be pushed up against the coop. It will need to be secured with some mechanical fasteners, but there is nothing too tricky about that.

36 Add the finishing touches. This photo shows the final configuration of the coop's interior. You'll notice that I moved the chicken ladder (M) to the left-hand side of the coop, and this necessitated moving the nest box—no biggie, since it was built as a separate component. I also cut a 14" x 14" (355 x 355mm) hole in the floor of the coop and installed a second chicken ladder (M) so that the birds can easily move from the coop to the run. Finally, I put a pair of roosts in the back of the coop so that the birds have a nice place to sleep. The bars of the roosts (V) are screwed onto a pair of blocks (W) I screwed directly to the walls of the coop.

36

KEEPING CHICKENS IN THE FAMILY

The joy of introducing young children to animals can't be overstated. Our daughter, Abigail, has had a ball with these chicks since day one.

 The average chicken egg is 70 to 90 calories.

SIMPLY SALVAGED
Rustic Board-and-Batten Siding Complements This 3-Level

This structure was built with the same "work with what you've got" ethic that I bring to many of my coops, and the design evolved in a fairly unique way, due in large part to the one-of-a-kind materials I used. I didn't happen to have a lot of scrap plywood on hand at the time, so rather than constructing large wall and roof panels with wide pieces of plywood as I normally might, I built one entire wall with a horizontal board-and-batten siding that turned into a rather eye-catching detail. I also had a pair of old windows in perfect shape, so I used them as doors—one for the run and one for the coop. This is not only a nifty way to recycle some old building materials, but it creates—in my opinion, at least—a fun and funky look. In the summer, I think it might be a good idea to replace the glass door with a screen door to allow better ventilation—your chickens will thank you for giving them a nice cool place to sleep at night.

Because the coop itself is tall with a small footprint, it might fit nicely into a yard where space is at a premium. And, even though its floor plan measures just 2½' x 3' (800 x 900mm), it is roomy enough for about six birds—the interior is divided into three floors that are connected by a series of chicken ladders.

 Plymouth Rocks are popular brown egg-layers and become tame easily.

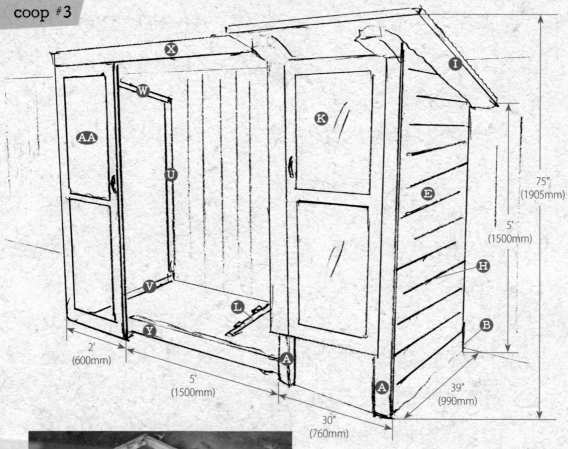

75"
(1905mm)

5'
(1500mm)

2'
(600mm)

5'
(1500mm)

39"
(990mm)

30"
(760mm)

1

2

1 **Frame the coop's right side.** I framed this coop with 2x4s. I laid out the side profile first, beginning with the front leg (A) and the back leg (B). The span between them is 39" (990mm). The lower stretcher (C) is easy to set into place, but the upper one (D) requires angled cuts. I set my miter saw to 13° and cut the tops of the legs and the back edge of the rafter to neatly accommodate the sloped roof. I joined the parts with 3" (75mm) all-weather deck screws.

2 **Build the second side.** To make sure that the sides are mirror images of each other, I simply built the second one on top of the first one. In this way, even if the first one is slightly out of square, it shouldn't cause any problems later because any discrepancies will be repeated in the second one and everything, therefore, should line up fine. A decade ago I coined a phrase to describe this concept: "It is better to be consistently wrong than occasionally right." This is not meant to excuse sloppy craftsmanship, but in reality, mistakes happen, and you can often work around them if you're clever.

MATERIALS LIST

item	material	dimensions	quantity
Ⓐ Front leg	2x4 (38 x 89mm)	6' (1800mm)	2
Ⓑ Back leg	2x4 (38 x 89mm)	5' (1500mm)	2
Ⓒ Lower side stretcher	2x4 (38 x 89mm)	39" (990mm)	2
Ⓓ Upper side stretcher	2x4 (38 x 89mm)	45" (1145mm) (approximate)	2
Ⓔ Siding	⅝" x 6" (16mm x 150mm) pickets	3½' (1100mm)	12 (per side)
Ⓕ Back panel	¾" (17mm) OSB or plywood	2½' x 4' (800 x 1200mm)	1
Ⓖ Bottom panel	¾" (17mm) plywood	2½' x 3½' (800 x 1100mm)	1
Ⓗ Battens (for trim)	¾" x 1" (17mm x 25mm) pine strips	3½' (1100mm)	13 (per side)
Ⓘ Roof	¾" (17mm) plywood or similar	3½' x 4' (1100 x 1200mm)	1
Ⓙ Left side trim	¾" x 1½" (17 x 40mm) stock	18", 39", 12" (458, 990, 305mm)	2, 2, 3
Ⓚ Door	Reclaimed wood	5' x 22" (1500 x 560mm)	1
Ⓛ Chicken ladders	Pine scraps	3', 2', and 1' (900, 600, 300mm) long	3 total
Ⓜ Cleats	1x2 (25mm x 51mm) pine	2' (600mm)	5
Ⓝ Vertical platform post	2x2 (38 x 38mm)	18" (458mm)	1
Ⓞ Interior bottom platform	¾" (17mm) plywood	2½' x 22" (800 x 560mm)	1
Ⓟ Interior upper platform	¾" (17mm) plywood	2½' x 22" (800 x 560mm)	1
Ⓠ Nesting box panels	¾" (17mm) plywood	1' x 1' (300 x 300mm), 1' x 38¾" (300 x 1000mm)	4, 1
Ⓡ Roost bars	1¼" (32mm)-diameter closet rod	30" (762mm)	2
Ⓢ Roost sides	⅝" (15mm) plywood	14" x 6" (355 x 152mm)	2

RUN (see page 136):

item	material	dimensions	quantity
Ⓣ Front leg	2x4 (38 x 89mm)	6' (1800mm)	1
Ⓤ Back leg	2x4 (38 x 89mm)	5' (1500mm)	1
Ⓥ Lower side stretcher	2x4 (38 x 89mm)	39" (990mm)	1
Ⓦ Upper side stretcher	2x4 (38 x 89mm)	45" (1142mm) (approximate)	1
Ⓧ Upper stretcher	1x6 (19 x 89mm)	6' (1800mm)	1
Ⓨ Lower stretcher	1x6 (19 x 89mm)	4' (1200mm)	1
Ⓩ Front vertical support	2x4 (38 x 89mm)	6' (1800mm)	1
ⒶⒶ Door	Reclaimed window or plywood	6' x 2' (1800 x 600mm)	1
Screws			
Paint			
Chicken wire			

 One egg contains 6g of protein, or 12% of the recommended daily value.

3

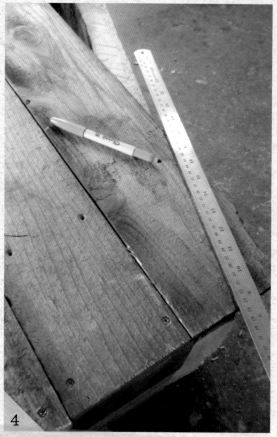

4

3 Attach siding to the right side. I used ⅝" (16mm) fence pickets (E) (recycled, of course) as siding for the coop. I attached them with screws, and didn't worry about small gaps of ¹⁄₁₆" (2mm) or ⅛" (3mm) between the slats—we'll cover them up later.

*Near the end of the project (Step 23), I decided to put siding the whole way to the bottom. If you prefer that look to chicken wire, might as well put it on now.

4 Mark the roofline. It is easy to apply siding where the roof tapers if you use this trick: just set the siding in place and then, while standing directly above it, use a ruler to draw a line where the top of the 2x4 runs. When the excess has been removed, the piece will fit just fine.

5 Trim the excess siding. Here's a shot of the siding, trimmed to fit. After it has been screwed on, I flipped the completed side over and applied siding to the other side panel.

DESIGN CONSIDERATION: MEASURING A COOP

I've never had a discussion with an official inspector— in our area, I suppose that would be someone from the Salt Lake County Animal Services Department—but common sense seems to indicate that, when you've got a tall coop with more than one level inside, you could add up the square footage of each floor to get an accurate idea of the coop's capacity. So, in this case, the footprint of the coop is just 2' x 3.5' (600 x 1100mm), which is 7 square feet (.7 square meters), or enough space for three chickens, according to our local requirement of 2 square feet (.2 square meters) per bird. However, the total floor space is 16 square feet (1.5 square meters), which would allow eight. My opinion is that this might be a bit cramped for eight birds, but six would fit fine. In any event, if you have a small lot, a tall coop may be the most space-efficient way to house your chickens.

5

6 Install siding on the left side. On the left side, I used plywood for siding, because I had some on hand. However, you could continue using fence planks (E). My overall plan for this coop was to have the chickens come and go through a small door on the left-hand side, so I used a framing square to mark out the door. 1' x 1' (300 x300mm) is sufficient.

7 Cut the doorway. I cut out the doorway with a jigsaw in a minute or so.

8 Paint the left side. Because I planned to apply some trim to this side of the coop, and I wanted the trim to be left natural in color, I painted the side prior to attaching the trim. This is way faster and easier than trying to get a precise result by trimming it out and then painting it.

9 Attach the back panel. With the two side panels completed, I tilted them up in the air and screwed on a back panel (F).

Most laying hens in the U.S. are Leghorns.

10

11

12

13

10 Attach the floor. After you've attached the back, put the structure upright. This provides good access to the interior of the coop, so the logical next step is to put in a floor (G). Notch the floor panel to fit around the legs and rest securely on the 2x4s.

11 Attach the battens. To create an interesting look, I covered the joints between the pieces of siding with 1" (25mm) wide battens (H). This creates a more airtight assembly, as well.

12 Attach the roof panel. Because the total width of the coop is only 30" (762mm), and such strong timbers were used as rafters (D), it is easy to build a strong roof (I). I used an old hollow-core door, but plywood would also work.

13 Add trim to the left side. When the paint had dried, I trimmed out the left side of the coop (J). This is mostly aesthetic, but the trim around the doorway also provides a place to anchor some door hinges if you wish to add a door.

14 Make the door and attach it. I made the door (K) out of an old window. I thought it looked neat, and I couldn't think of a reason not to. You can use whatever you have that fits the necessary size.

15 Build the chicken ladder. A standard-issue chicken ladder (L) will make it easy for the residents to come and go as they please.

16 Install the cleats. The interior of this coop has three levels that the chickens can hang out on. Each floor is supported by a set of cleats (M) screwed to the walls of the coop. I recommend putting the first set of cleats 18" (458mm) from the bottom, and the second set at 30" (762mm).

Egyptian records show that fowl were used for eggs in 1400 B.C.

17 Install the bottom platform.
This panel (O) gets notched to fit around the posts in back—an easy task with a jigsaw or a sharp handsaw. I also supported the left front corner with a vertical 2x2 (N), because screwing a cleat along the left-hand side wasn't a very good option—the siding wasn't thick enough to hold a screw.

18 Install the upper platform.
Construction in this kind of situation goes one step at a time: after the first level was in place, I could see where to position the next one (P). In this case, across from the first level (O) worked best.

19 Build the nest boxes. I decided to do a row of three nest boxes (Q) on the bottom floor. The top side of the nest boxes also provides a place for chickens to walk around, so there really isn't a loss of overall usable floor space.

17

18

19

20

21

22

20 Build the interior chicken ladders. A series of chicken ladders (L)—custom-sized for each spot—will help the future residents hop around between the different levels.

21 Make the roost. Because it is built as a freestanding unit, this roost can be removed to make cleanup a cinch. The two 30" (762mm)-long bars (R), screwed into plywood sides (S), provide 5' (1500mm) of total roosting space—enough for six chickens if you apply a standard of 10" (255mm) per bird.

22 Position the roost. I put the roosts at the top of the coop—my thinking was that this might tend to be the warmest spot in the winter, and in the summer, I figured that the birds might prefer roosting outdoors.

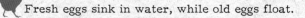 Fresh eggs sink in water, while old eggs float.

23 Complete the coop. I decided to run the siding (E) all the way down to the bottom on the right-hand side of the coop. I had intended to cover this area with chicken wire, but I had some extra siding on hand, and I thought that it might look good this way. I built a run attached to the coop—see page 136. I also decided on a funky fuchsia color for some of the elements of the structure to add a little character. I didn't want to cover up all of the natural wood tones, though, because I think it actually looks better to have a bit of contrast. Add stamps, shape cut-outs, and whatever else you'd like!

23

POPULAR BREEDS FOR BACKYARD CHICKEN FLOCKS

The following breeds are highlighted here because they possess great all-around characteristics: They are friendly, hardy in all climates, and produce at least three eggs per week, if not four or five. The majority (exception being Leghorn) are also dual purpose, meaning they are ideal if you wish to harvest eggs as well as meat. Most are average or less-than-average brooders (exceptions being New Hampshires, Orpingtons, and Wyandottes), meaning they won't try overly hard to sit on their eggs—though keep in mind that broody breeds are best if you wish to hatch chicks. This list is a good general place to start, but consult the Internet and chicken breed books for more information—there are hundreds of breeds to choose from.

Ameraucana
Medium-sized blue or green eggs.

Australorp
Large brown eggs.

Brahma
Light brown medium-size eggs.

Delaware
Large brown eggs.

Leghorn
Egg layer, flighty and active, large white eggs.

Maran
Large dark brown (chocolate) eggs.

New Hampshire Red
Frequently broody, large brown eggs.

Orpington
Frequently broody, large light brown eggs.

Plymouth Rock
Large light brown eggs.

Rhode Island Red
Large brown eggs.

Sussex
Medium-sized brown eggs.

Wyandotte
Frequently broody, large brown eggs.

JIMMY & BRIT

Salt Lake City, Utah

Jimmy Ruff and Brit Merrill became chicken owners about three years ago, and they've loved just about every minute of it. Their most important influence was their neighbor Rick, who is locally known by the moniker "Johnny Chickenseed": he's a gregarious guy who loves turning people onto the fun of backyard chickens. Brit recalled that they had spent a lot of time at his house—he often hosts large potluck Sunday dinners—and got hooked on the entertainment value that Rick's birds provided. Jimmy was less enthusiastic, but he signed on for the project and quickly became a convert. He notes that getting chickens really wasn't a quantum leap from what they were doing, but just the next logical step. They were already committed organic gardeners, after all, and the idea of producing even more of their own food was pretty appealing.

building the coop

One of their biggest challenges was in building the coop itself, and they were a bit humbled by the task. Brit laughs now when she looks back and remembers thinking it wouldn't take more than a day. They had never built anything before, so it was a big project for them, but they are both thrilled with the results—even

← Jimmy and Brit have a neat setup. It consists of a tall home-built coop situated within an enclosed run. It is a spacious area—about 15' x 15' (4600 x 4600mm), or 225 square feet (21 square meters), which could provide enough space for up to 22 chickens. At the moment, they only have five birds, though, and there's nothing wrong with some extra elbow room.

↑ Chickens love having places to hide out, and these old boards leaned up against the fence provide some highly desirable nooks and crannies. They also create some shade, which is essential on hot days.

⬆ Jimmy and Brit take great pleasure in raising a flock of five chickens in their backyard. They choose breeds based on what birds they think are most beautiful, rather than egg production, and say they still end up with plenty of eggs.

Huevos Rancheros, Mexican: Fried eggs served over tortillas with tomatoes, chiles, refried beans, and avocado.

though it took them a bit longer than they had anticipated. Jimmy recalls that they got it done over the course of two weekends. They began the process by looking at designs on the Internet and checking out some other coops in town; then they came up with their own version. Because most of their materials were secondhand, they did have to work within those constraints, but that wasn't a problem. Jimmy says that if he were to do anything different, he would make the coop a bit deeper, but the current coop performs just fine. They're quite proud of it, and note that it has good airflow, conveniently situated nest boxes, and a person-sized door for good access.

entertainment value

One of Jimmy and Brit's favorite benefits of having chickens comes from watching the birds go about their daily routines. Jimmy is a biologist—an animal behaviorist, to be precise—and he finds a lot of joy in watching the birds interact. He mentioned that we use the term "pecking order" to describe human relationships, but that it developed in the context of flocks of birds, and he marvels at the social structures that his chickens seem to have. Brit enjoys having a cup of tea and watching the chickens roam around—she says it is a really calming experience for her.

chicken poop fertilizer

In addition to the emotional satisfaction that Jimmy and Brit get from their flock, they also have found an exceptional advantage in the quality of compost that it produces. Jimmy says that raising the chickens for eggs alone doesn't break even

The coop can be easily closed up by raising this combination door/chicken ladder.

Nest boxes don't have to be anything fancy—this spot is clean, dry, and full of soft bedding.

← Because Brit and Jimmy are active gardeners, they set up their compost pile right next to the chicken coop: this makes it really convenient when it comes time to clean out the coop.

↓ The top of the coop is open to allow plenty of airflow. Remember, good ventilation correlates with healthy chickens.

financially (compared to store-bought eggs), but when they factor in the savings on buying good organic compost for the vegetable garden, they actually come out ahead of the game. They said that it took them a little while to develop a system for composting with the guano that their birds produce, but that they've gotten it down to a science now. They've learned to put the manure onto their compost pile and then wait six to nine months for it to "cool off," as fresh chicken manure is too nitrogen-rich to be used right away—it would actually burn plants and roots. This makes a great closed loop on their property: they feed the chickens, and the chickens feed the garden, which feeds the people.

← The interior of the coop boasts several "natural" roosts made from old tree branches.

 During the colder months, try wrapping the run with plastic to insulate and keep snow out.

community builder

Jimmy and Brit have also discovered that owning chickens is a great community builder. They say their neighbors often bring their young children over to visit the chickens, and they are proud to be able to provide this fun and educational experience. They note that the question, "How are the ladies doing?" comes up pretty early in a lot of their conversations with friends. They've also met a lot of fellow chicken owners in the area that they might not have met without this common interest.

tips

Their experiences have taught them a lot about the day-to-day business of raising chickens, and both Brit and Jimmy were generous in sharing a lot of tips for people who are thinking about getting into the game. I'll share a few of them here:

- **Do some research into chicken breeds.** Jimmy has observed that most chickens fall into one of the following categories: layers, meat birds, and fancy birds. His advice is to find the chickens that appeal to you the most, because the difference between top-notch layers and just regular ones really isn't a make-or-break factor for a home flock. In the long run, getting five or six eggs a week from a particular bird probably doesn't matter; he and Brit choose birds they think are beautiful, and they always end up with plenty of eggs. They also mentioned they try to avoid the breeds that are more likely to go broody, meaning that they will be compelled to sit on the eggs to try to hatch them. If you don't have a rooster around, obviously the egg will not be fertilized and cannot hatch—but while

← Some of the ladies are enjoying a mid-afternoon watermelon snack. All birds should be so lucky!

the hen is sitting on the egg, she is not contributing to daily egg production, and can also cause a stir by hogging a nest box for days or weeks. They both also take pleasure in raising endangered and heritage breeds and encourage others to do the same.

- **Watching your chickens is important to their health.** Brit and Jimmy enjoy spending time in their backyard with their chickens, but this is more than just a fun activity: by getting a feel for how the chickens behave on a regular basis, their owners are more in tune with how the birds are doing and can easily tell if a chicken isn't doing well or may need some extra attention. They have a chicken named Leia that suddenly began acting differently, and so they medicated her with over-the-counter veterinary antibiotics. This produced a major improvement within twenty-four hours, and both Jimmy and Brit feel that Leia may not have survived without the intervention.

- **Start with birds that are the same age and/or size.** Brit mentioned the challenges that can arise when birds are brought from disparate situations into a new flock—the results can be downright disturbing as chickens peck at each other and sort out their social hierarchy. To avoid this, she prefers having a cohesive flock of birds that were raised together rather than introducing new birds that might struggle to fit in.

- **Set up multiple food and water stations.** Experience has taught Jimmy and Brit to keep two water stations filled up, because they've come home on more than one hot day to find a water station knocked over. They've also noticed dominant birds guarding the water and preventing birds of lower status from drinking. Setting up two watering stations ensures that everybody gets enough to drink. This is particularly important when bringing new birds into the flock.

- **Have a rooster contingency plan.** While many (perhaps most) people don't want or need roosters, and chickens are sold according to sex, the process isn't foolproof, and it is not impossible to purchase a baby chick that grows up into a male instead of the hoped-for female. Jimmy says that the accuracy rates are about 80-90%, and while these are pretty comfortable odds, they're not perfect. Knowing what you'll do if you end up with a rooster is a good idea, he suggests, because then you'll be prepared when and if that time comes. Are you interested in eating it? Is there someone else around who will take it off your hands? Having a plan never hurts.

Be sure to select breeds that do well in your climate.

RUSTIC SOPHISTICATION

Fish Scale Shingles and a Front Porch Add Victorian Class

This coop was built as a prize for a drawing sponsored by Wasatch Community Gardens as part of their annual Tour de Coops. For this reason, I thought it would be fun to dress it up a bit with some fancy trim. The fish scale shingle work below the gable on the front of the coop was inspired by the ornate exteriors of many of the older homes in our area, although I had to scale the elements down a bit in size. This Victorian-inspired detailing was fairly time-consuming, but completely worth it. I also embellished the front of the coop with a porch, which helps the coop to feel a bit less utilitarian and more sophisticated, as much as a chicken coop can be sophisticated. The decorative columns that hold up the porch roof are reclaimed spindles from an old staircase.

In terms of practicality, this coop offers great nest box accessibility through a door on the side, and as always, the nest boxes and roosts are removable to facilitate cleanup. The materials for the coop were 100% recycled—even the shingles were purchased secondhand from a neighbor who ended up with some extras that were too good to throw away.

 If any leg scales are sticking out, rub them with mineral oils–this will kill the mites.

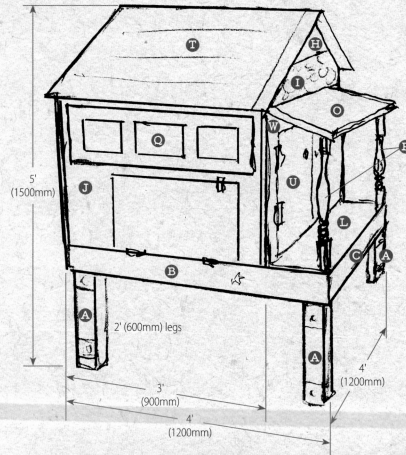

5'
(1500mm)

2' (600mm) legs

3'
(900mm)

4'
(1200mm)

4'
(1200mm)

1 Cut the legs to length. Most of the coops I build are raised off of the ground on legs, but I do use a couple of different approaches in terms of the construction. In this case, I used a set of 4x4 cutoffs (A) scavenged from a jobsite. You could also use a single 8' (2400mm)-long 4x4 and cut it into sections.

2 Connect the legs and stretchers. I put the 4x4s together with a series of horizontal stretchers. The side stretchers (B) are 38" (965mm) and the front/back stretchers (C) are 4' (1200mm). A project like this doesn't require fancy joinery—just lots of 3" (75mm) all-weather screws. Be sure to install the side stretchers (B) on the inside faces of the legs. This will keep the total depth of the coop just the right size to fit the 48" (1220mm) sheet of plywood.

MATERIALS LIST

item	material	dimensions	quantity
Ⓐ Legs	4x4 (89 x 89mm) square posts	2' (600mm)	4
Ⓑ Side stretchers	2x4 (38 x 89mm)	38" (965mm)	2
Ⓒ Front/back stretchers	2x4 (38 x 89mm)	4' (1200mm)	2
Ⓓ Platform (floor)	⅝" (15mm) plywood	4' x 4' (1200 x 1200mm)	1
Ⓔ Left/right side-wall framing	2x2 (38 x 38mm)	3' (900mm)	4
Ⓕ Top/bottom side-wall framing	2x2 (38 x 38mm)	33" (838mm)	4
Ⓖ Corner plates	¼" (6mm) plywood	5" x 5" (128 x 128mm)	8
Ⓗ Gable end panels	¼" (6mm) plywood	2' x 4' (600 x 1200mm)	2
Ⓘ Shingle blanks	¼" (6mm) plywood	6" x 3" (152 x 75mm)	50
Ⓙ Side wall panel skins	⅜" (8mm) plywood	3' x 3' (900 x 900mm)	2
Ⓚ Back panel	⅝" (15mm) plywood	4' x 3' (1200 x 900mm)	1
Ⓛ Front porch platform	¾" (17mm) plywood	4' x 1' (1200 x 300mm)	1
Ⓜ Front porch edging	¾" (19mm) pine	6½' (2000mm) (trimmed to fit)	1
Ⓝ Cleat for porch roof	¾" (19mm) pine	4' x 1½" (1200 x 320mm)	1
Ⓞ Porch roof	¼" (6mm) plywood	4' x 15" (1200 x 380mm)	1
Ⓟ Porch supports	Spindles	3' (900mm)	2
Ⓠ Window	Repurposed cabinet door with panes, or build from 2x4s	32" x 1' (81 x 300mm)	1
Ⓡ Back gable support	¾" (19mm) OSB	44" x 6" (1118 x 152mm)	1
Ⓢ Rafter	2x4 (38 x 89mm)	3' (75mm)	1
Ⓣ Roof panels	⅜" (8mm) plywood	3½' x 2½' (1100 x 800mm)	2
Ⓤ Front doors	Repurposed cabinet doors, or plywood	26" x 20" (660 x 508mm)	2
Ⓥ Front door support	¾" (17mm) plywood	33" x 3½" (838 x 90mm)	1
Ⓦ Panel above front door	½" (11mm) plywood	40" x 2½" (1015 x 65mm)	1
Ⓧ Blank for sunburst	¼" (6mm) plywood	4' x 16" (1200 x 405mm)	1
Ⓨ Nest box sides	¾" (17mm) plywood	1' x 1' (300 x 300mm)	3
Ⓩ Nest box top	¾" (17mm) plywood	34" x 1' (865 x 300mm)	1
ⒶⒶ Roosting bars	¾" (19mm) pine	30" x 1" (762 x 25mm)	2
ⒷⒷ Roost base	¾" (17mm) plywood	14" x 4" (355 x 102mm)	2
Screws			
Glue			
Chicken wire			
Hinges			
Paint			

 Eggshells are permeable and draw in air to create an air cell.

3

4

3 Install the floor platform. Screw on the floor platform (D). The base for this coop resembles a table of sorts.

4 Construct the side walls. I decided to frame the side walls on the floor and then attach them to the base. This method made it easy to assemble the parts without an assistant. To ensure a strong connection between the 2x2 "studs" (E, F), I screwed on square plates (G) made from scrap ¼" (6mm) plywood. Verify that the wall panels are square—this kind of thing pays off later by helping the entire coop go together evenly.

5 Attach the side wall panel frames to the base. I used 3" (75mm)-long screws so that the walls would be securely connected to the beefy framing below the base's surface. You could also apply wall sheathing prior to securing the walls to the base, but I find it easier to screw the frames down when there is no sheathing in place.

5

GET FEED TO THE BIRDS

Hanging a feeder up in the air a bit is a great way to keep mice from getting into the chicken feed. Storing larger quantities in plastic or metal containers with lids is another important part of that equation.

You can see that I set the wall framing in from the edge about ¾" (20mm)—this is to provide a bearing surface for the wall surface to sit on. It probably isn't necessary, but that was how I went about it.

6 Complete the side wall attachment. Here's a glimpse of the progress with the side wall framing in position.

7 Attach the front gable panel. With the side walls framed up, the next step is to connect them with a gable end panel (H). This lends considerable strength to the structure. The end panel isn't a triangle, exactly—I made the panel's bottom extend 4" (102mm) to provide a place to screw the panel to the framing. A set of inexpensive spring clamps held the whole thing together while I worked.

8 Create shingle blanks. One of the distinguishing features of this design is the shingle pattern on the front gable panel. This particular pattern requires two types of shingles: half-rounds and arrows. Start by cutting out a pile of 6" x 3" (152 x 75mm) blanks (I), and then make a template for the half-rounds. A compass set to a 3" (75mm) radius makes cutting all 50 easy.

Egg Drop Soup, Chinese: Whisk a stream of beaten eggs into boiling chicken broth; add green onions.

9

10

9 Cut out the half-rounds. I cut the shingles out on the band saw, and to save time, I stacked up eight or nine blanks and held them together with tape. This speeds things up considerably.

10 Cut out the arrows. The arrows are cut out in the same way. Use a compass set to the same diameter and draw two arcs with the compass point set at the bottom corners of each blank.

11 Lay out the shingles. The pattern is created by laying out rows of half-rounds and arrows, with the arrows shifted 1½" (38mm) to the side.

12 Start attaching shingles to the gable. To apply the shingles to the coop, I drew a level line 2" (50mm) up the gable panel and used it to align the bottoms of the half-rounds.

11

12

13

CHOCOLATE EGGS?

Photo courtesy of Abrahami.

Marans lay dark brown eggs, also called chocolate eggs. The shell is much easier to see and remove when peeling hard-boiled eggs.

14

13 Finish applying shingles. When you have two rows of half-rounds, a diamond is created between them. Pretty neat. You can also see that I didn't worry about trimming the shingles on the end—I just let them hang proud, as I knew I could trim them later on.

14 Trim the overhanging shingles. You can save a lot of time and get a great result by simply scribing a line across the front of the shingles and cutting away the excess with a jigsaw.

 To candle an egg, hold it up to a bright light and examine the contents.

15

16

17

15 **Attach plywood to the side walls.** The wall panels stiffen up with the addition of ⅜" (8mm) plywood skins (J). Attach them with glue and screws.

16 **Attach the back panel.** The back panel (K) goes on in the same way, and after it has been added, the whole structure becomes quite strong.

17 **Develop the front porch.** This design features a front porch where the hens can hang out. I laid down a ¾" (17mm)-thick plywood (L) to help define this area. I then trimmed out the edges with some mitered trim (M) to create a more finished-looking edge.

18 **Rip the roof porch cleat.** The front porch is covered by a sloping shed-style roof. The roof panel (O) is attached to the coop by an angled cleat (N). I ripped it at a 15° angle on my table saw so one long side has the appropriate angle. Attach it flush to the bottom of the front gable (H) with screws.

18

19

20

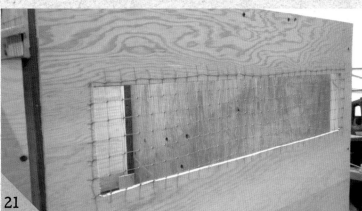

21

19 Attach the porch roof. The roof should provide some nice shade for the residents.

20 Install the porch supports. To support the roof panel, I used a pair of stair balusters (P). Fitting them into place was easy—I drew a line to indicate the intersection of the roofline and baluster, and cut away the excess on my chop saw. The angle is a given—15°—just like it was at the top edge of the roof cleat.

21 Cut in a side window. I like to build extra window openings to allow for healthy cross-ventilation. I had a cabinet door with panes that I thought would make a nice window (Q). I cut out an opening to fit it and stapled a layer of chicken wire over the top. If you don't have a cabinet door like this, build one out of 2x4s or just leave the window screened and plain.

22 Attach the window covering. I debated how to attach it—hinges along the top or bottom edge seemed like a natural choice—but ultimately I decided to screw it down.

22

 Feed your chickens limestone to ensure their eggshells have enough calcium.

23 Attach the gable support. The back wall of the coop needed a gable end panel (H) to help support the roof. To hold the gable in place, I screwed a scrap piece of OSB (R) onto the back wall with a few inches of surface area protruding above its top edge.

24 Screw in the gable panel. The back gable panel (H) is a mirror image of the one at the front—minus the fancy trim, that is.

25 Install the rafter. I used a 2x4 (S) to connect the gables (H) at their peaks. This provides a nice place to fasten the roof panels. This roof system is very simple, but the spans are quite small, and the roof pitch is steep enough to shed snow readily.

26 Attach the roof panels. The roof panels (T) are ⅜" (8mm) plywood. The total unsupported span is only 3' (900mm) across, and as long as there are plenty of screws sunk into the 2x4 (S) and 2x2 (F) framing at the top and bottom edges of the roof, I can't imagine that it will flex one bit.

23

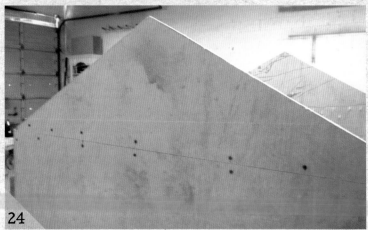

24

25

26

27

28

27 Attach the front door supports. With the roof (T) in place, I turned my attention to attaching the doors (U). I had an old set of kitchen cabinet doors. To make their odd size fit, I attached vertical pieces of ¾" (17mm) plywood (V) to the edges of the door opening. This also gave me a sturdy place to screw down the hinges.

28 Attach the front doors. I hadn't thought of this until I was under way, but in retrospect, it is obvious that the sloped roof would prevent the doors from opening if they extended all the way up. This meant that I had to trim the doors so they were short enough to clear the front edge of the roof. This worked, but it created an approximately 4" (102mm) gap above the doors.

29 Cover the gap. Every problem has a solution—I covered up the gap with a 4" (102mm)-wide scrap (W).

EGGNOG

English and American beverage

6 eggs
3 cups milk
7 Tbsp granulated sugar

1 Tbsp vanilla extract
½ tsp ground nutmeg
½ tsp ground cinnamon

Beat the eggs thoroughly. Add the milk and sugar; beat until the mixture thickens. Add vanilla and spices. To make this drink alcoholic, add rum, whiskey, or brandy to taste.

29

Sometimes hens will eat their own eggs. To prevent this, collect eggs often!

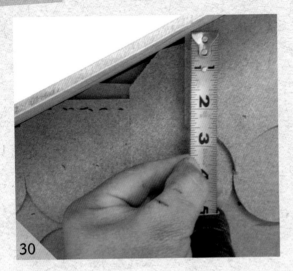

30

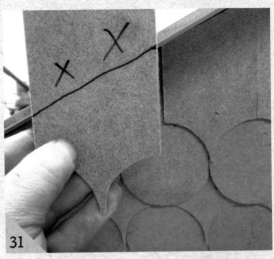

31

30 Mark the last few shingles.
Because I hadn't finished all of the shingle trim (I) prior to attaching the roof (T), I had to use a different strategy for fitting the last few pieces into place. The method I used was straightforward—I measured the amount of vertical run on the adjacent shingle (in this case it was 2½" [65mm]), and used that measurement to lay out the cut on the next shingle.

31 Cut the remaining shingles and attach them. I put a mark 2½" (65mm) up on the right-hand side of the next shingle, and then I held it out along the front edge of the roof. I made sure it was plumb by looking past it to see that its side was parallel to the adjacent shingle. When I felt it was aligned, I drew a diagonal line from right to left that mimicked the slope of the roofline. I cut away the excess on the band saw and it fit nicely. You may need to try this a couple of times to get a feel for it, but if you have a good eye, this technique should work just fine.

32

32 Paint the front gable. I had envisioned a small sunrise motif (X) above the shingle trim. The sunrise motif requires a back panel (the front gable [H]), which I painted blue.

33 **Make and attach the sunrise overlay.** I detailed the construction of a sunrise overlay on page 24, so I won't overdo it here. The overlay (X) went on smoothly and easily.

34 **Create side doors.** While the addition of the front porch is a neat effect, it might make it harder to reach into the coop to collect eggs. I decided to build a door on the side to provide easy access. I used a jigsaw to plunge-cut along the outline for the door. The cut-out was perfectly usable: all I had to do was add hinges and a latch. This is probably the easiest way to make a door.

POACH AN EGG

Oil a metal ladle and insert it into boiling water until it's almost covered with water. Crack an egg into the ladle and let it cook a bit—allow a little of the boiling water into the ladle and cook for about 3 minutes.

Frittata, Italian: Sautéed veggies mixed with egg and topped with cheese, served in pie slices.

35 Build the nest boxes. The size of this coop—3' x 4' (900 x 1200mm), or 12 square feet (1.1 square meters)—dictates that it could comfortably house up to 6 chickens, so I put in two nest boxes to maintain a ratio of one nest box for three birds. I built the boxes by screwing and gluing three sides (Y) to one long top (Z).

36 Build the roosts. These roosts are a fairly standard design for me—they're quick to make, they're sturdy, and they're removable to make cleanup a snap. Just screw the roosts (AA) to the supports (BB). The parallel roosts offer a total of 75" (1905mm) of roosting space, which means there is just over 12" (305mm) of room for each bird—just a bit more than the recommended 8" to 10" (205 to 255mm) that most experts suggest.

35

36

37

37 Add the final touches. With the coop nearly completed, I decided to focus on the aesthetics, which mostly meant coming up with a creative painting scheme. Sometimes I do this by trial and error, and end up repainting parts of a coop a couple of times to get the right look and feel. You have a lot of options for putting on a roof—as long as it is watertight and durable, you'll be in good shape.

 Eggs have an invisible protective coating, called bloom, to keep out bacteria.

LITTLE BIG BARN
A Low Design to Create Easy Access for Children

The story behind this coop is a fun one—a young family in a small town was intrigued by the idea of raising chickens, and they asked for my help in planning their new venture. They figured on having 10-15 chickens, which helped to determine the size of the coop they'd need, and they had already identified the perfect spot: a little-used side yard that measured 20' x 15' (6100 x 4600mm). This meant that they had a 300 square foot (27.9 square meter) run, which would be big enough for 30 birds; extra room to roam is always a good thing. Another advantage to their proposed location was the fact that it was fenced in with 6' (1800mm) vinyl fencing on two sides, and it was bordered on a third side by the house itself. This meant that we would only need to enclose one side of the whole area, which would save time, money, and materials.

The coop itself is fairly straightforward. It measures 7' x 4' (2100 x 1200mm), or 28 square feet (2.6 square meters), which would comfortably house up to 14 chickens using a standard of 2 square feet (.2 square meters) per bird. I kept the interior open so the movable roosts and nest boxes could be configured in whatever way seemed easiest. Two sets of double doors provide great access

for cleaning and plenty of ventilation. Speaking of ventilation, a large clerestory window at the top of the coop allows for additional airflow. I provided a cover that could be screwed on during cold weather. I also made sure to keep the base low to the ground—the legs are 16" (405mm) high—because the coop needed to be within reach of the children who would be tending the chickens.

Flan, Spanish: An egg custard dessert topped with caramel sauce.

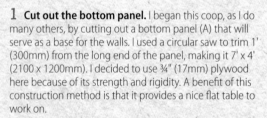

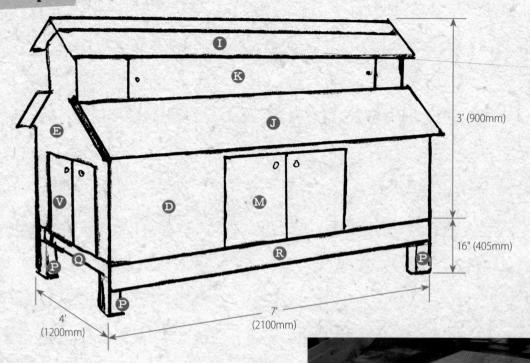

1

2

1 **Cut out the bottom panel.** I began this coop, as I do many others, by cutting out a bottom panel (A) that will serve as a base for the walls. I used a circular saw to trim 1' (300mm) from the long end of the panel, making it 7' x 4' (2100 x 1200mm). I decided to use ¾" (17mm) plywood here because of its strength and rigidity. A benefit of this construction method is that it provides a nice flat table to work on.

2 **Frame the long walls.** I framed the walls with 2x3s (B, C), joined together with long screws. I captured the vertical members (C) between the top and bottom boards (B)—this will ensure the force transferred from the roof (in terms of snow load, mainly) will not put undue pressure on the 2x3 on top and cause it to collapse. A pair of pipe clamps helped to keep the joints aligned while I worked.

MATERIALS LIST

	item	material	dimensions	quantity
A	Platform (floor)	¾" (17mm) plywood	7' x 4' (2100 x 1200mm)	1
B	Long wall top and bottom framing	2x3 (38 x 63mm)	7' (2100mm)	4
C	Long wall vertical framing	2x3 (38 x 63mm)	15" (380mm)	4
D	Long wall panels	¼" (6mm) plywood	7' x 1½' (2100 x 500mm)	2
E	End wall panels	¼" (6mm) plywood	4' x 3' (1200 x 900mm)	2
F	End wall supports	2x3 (38 x 63mm)	43" (1092mm)	2
G	Rafters (struts)	2x4 (38 x 89mm)	83½" (2120mm)	4
H	Roof panel cleats	2x4 (38 x 89mm)	1' (300mm)	4
I	Upper roof panels	¾" (17mm) plywood	7½' x 1' (2300 x 300mm)	2
J	Lower roof panels	¾" (17mm) plywood	7½' x 1½' (2300 x 500mm)	2
K	Clerestory side panels	¾" (17mm) plywood	7' x 1' (2100 x 300mm)	2
L	Door trimmers to reinforce door openings	2x3 (38 x 63mm)	11' (3400mm) total	1
M	Long wall doors	Repurposed cabinet doors, or ½" (11mm) plywood	30" x 16" (762 x 405mm)	2
N	Asphalt roofing shingles			1 bundle
O	Angled trim	2x4 (38 x 89mm)	7½' (2300mm)	2
P	Legs	4x4 (89 x 89mm) square posts	16" (405mm)	4
Q	Short side base stretchers	¾" (17mm) plywood	46½" (1180mm)	2
R	Long side base stretchers	¾" (17mm) plywood	7' (2100mm)	2
S	Nest box sides	¾" (17mm) plywood	1' x 1' (300 x 300mm)	5
T	Nest box top	¾" (17mm) plywood	6' x 1' (1800 x 300mm)	1
U	Roosting bars	¾" (17mm) pine	6' x 1' (1800 x 25mm)	2
V	Short wall doors	Repurposed cabinet doors, or ½" (11mm) plywood	27" x 14" (685 x 355mm)	2
	Screws			
	Construction adhesive			
	Chicken wire			
	Hinges			
	Paint			
	Felt roofing paper			
	Roofing nails			

 La Stracciatella, Italian: A soup with chicken broth, eggs, Parmesan cheese, semolina, parsley, and nutmeg.

3 Cut the long wall panels. The framing approach I use here is not unlike traditional house framing in that it relies on panels securely fastened to the frame's skeleton. The panels are known as shear walls; they add stiffness to the framing and prevent racking. I used ¼" (6mm) birch plywood for the long wall panels (D) and trimmed it to length with my jigsaw.

4 Attach the long wall panels. The completed wall assembly is quite strong. I used construction adhesive and lots of screws to hold the panels (D) and frames together.

5 Install the long walls. Once the first wall assembly is done, screw it into place from below. I ran a screw in every 1' (300mm).

6 **Review your work.** The long walls are built identically, and once they're installed, you can get a good feel for the overall proportion for the coop. It's a big one!

7 **Cut out the end panels.** I try to be creative with my designs, but in this case, I thought it would be fun to go for a traditional look, reminiscent of an old barn. I drew out a shape for the gable end panels (E) on ¼" (6mm) plywood and tweaked the angles and dimensions until it seemed just right. I didn't worry too much about the exact measurements involved—I just wanted it to look nice. This is hardly a scientific approach, but it works. After cutting out the first panel with a jigsaw, I used it as a template for the second one.

SANITARY PRECAUTIONS

Because we often have visitors who like to pet the chickens, we make sure to have some hand sanitizer around just be on the safe side. Is it necessary? Probably not, but it doesn't hurt.
For convenience, we came up with this simple holder. It consists of a loop of aluminum flashing that is screwed to a corner post on the outside of the run, at a height that even children can reach.

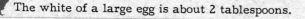
The white of a large egg is about 2 tablespoons.

8

8 Install end wall supports. To attach the gable end panels (E), I placed a length of 2x3 (F) on the floor of the coop in the gap between the side walls. I screwed it to the floor and side panel. This provided a place to screw the panel to.

9 Attach the end panels. The sides of the end panels (E) are simply screwed into the vertical 2x3s (C) of the adjacent walls.

10 Check the big picture. Astute readers will notice that the structure, although nice enough in appearance, lacks a few practical amenities—such as doors, to begin with. This will be covered soon enough! My goal is always to help shape up the "big picture" as quickly as possible, and it is much faster and easier to build components that are identical (i.e. pairs of walls) to begin with and then just modified as need be.

9

10

11 Install rafters. To securely attach the ¾" (17mm) plywood panels (I) for the upper roof, I added a series of struts (G) that ran with the long sides by screwing them into the end panels (E). This was the simplest way to build a strong roof while keeping whole project as light in weight as possible.

12 Identify the door locations. After the coop is framed up, it's easy to visualize the best locations and sizes for door openings. The sizes of the openings are often influenced by the range of old kitchen cabinet doors I have available. A framing square helps to lay out rectangular openings with ease.

13 Cut out the doorways. I used a jigsaw to cut out the doorway in just a couple of minutes. This is also a good time to screw the lower roof panels (J) to the struts.

14 Install roof cleats. Install cleats (H) at the top of the gable (E) to hold the upper set of roof panels (I). You will need to cut the 2x4s to fit the angle.

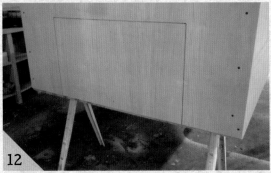

Egg Salad: Mix chopped hard-boiled eggs with mayo, mustard, relish, and celery salt.

15 Complete installation of clerestory panels. The side panels of the clerestory (K) are made of ¾" (17mm) plywood for strength and rigidity, and I attached them to the gable wall panels (E) with screws. I also screwed down the upper roof panels (I).

16 Cut a window into one clerestory panel. I decided to cut a large opening on one of the clerestory panels (K) to provide light and ventilation inside the coop. To keep predators out and chickens in, I stapled a layer of poultry netting across the back side of the panel. In the winter, I figured that the opening could be easily covered by a removable door.

17 Install the last clerestory panel. I'm used to working alone, but sometimes things go much faster if you have some help—my friend Ryan was around to lend a hand when it came time to hold this panel (K) in place.

EASY HOLLANDAISE SAUCE

French

4 Tbsp butter	1 Tbsp lemon juice
3 egg yolks	
2 Tbsp hot water	Salt and white pepper to taste

Melt the butter. Whisk the egg yolks; add the lemon juice while continuing to whisk. Slowly beat in the butter and water. Pour the mixture into a pot and whisk over low heat until the sauce thickens. Serve right away over steamed vegetables or eggs Benedict (see page 127).

15

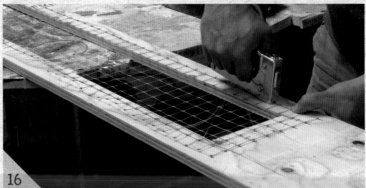

16

17

18

18 Decide on finishing details. Here's the basic structure—from here on, it could take shape in any way that you like in terms of paint colors, roofing materials, trim, doors, etc.

19 Reinforce the doorways. I decided to reinforce the edges of the door openings, both as a way of strengthening the walls as a whole, and also to provide a beefy place to sink screws for the hinges that would hold the doors on. I used 2x3s (L).

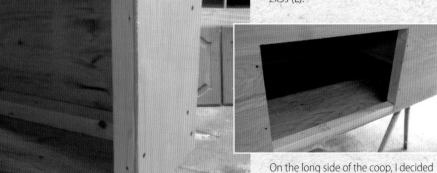

On the long side of the coop, I decided to leave the lower 2x3 in place in front of the door opening. It would've been easy enough to cut it out, but I figured that there was no harm in leaving it there—the chickens would have no problem walking right over it, and it does add a bit of beefiness to that part of the structure.

19

The best hard-boiling eggs are a week old—easier to peel because of the extra air that's leaked in.

20 Align the door hinges. Because these hinges will be visible from the outside of the coop, I like to make sure that they're all aligned consistently. The distance (2½" [65mm] from the door edge in this case) is arbitrary.

21 Paint the doors. I want to paint the trim elements (including the doors) white while doing the coop in a classic barn red. I've found that it is much easier to paint the trim prior to attaching it to the coop. This eliminates the need to mask off sections with tape, and it always makes for a neater result and a faster process.

22 Attach the doors and paint the coop. These doors (M, V) came with hinges pre-attached. This style of hinge is easy to use because they are screwed to the coop with the doors closed—this makes it a snap to position the doors correctly and then put in some screws without any guesswork. Paint the coop.

23 Install felt roofing paper. I decided to install a traditional asphalt shingle roof (N) on this coop, and the first step was to put down a layer of felt paper. I wrapped the edges of the roofing panels (I, J) to keep water from penetrating into the porous ends of the plywood. On a house, metal drip edges would be the solution.

20

21

22

23

24 **Complete application of tar paper.** Here's a glimpse of the upper roof, prepped for shingles.

25 **Begin nailing on shingles.** The shingles (N) went on pretty quickly—just be sure to use roofing nails, as their wide heads will help to keep the shingles fastened down without tearing through.

26 **Trim the shingles at the roof edges.** When laying down the shingles (N), each horizontal row should be offset from the row below it—this means that you'll need to cut off the ends occasionally. Here's an easy way to do it: just bend the shingle over the edge of the roof panel, and then bend it back and forth a few times.

27 **Trim the top row of shingles.** To make long cuts in shingles—whether for starter courses or at the top of the lower roof panels (you want to cut the shingle tabs off so it looks nice)—a straightedge and a utility knife get the job done easily.

24

25

26

27

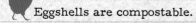
Eggshells are compostable.

28

29

30

28 Continue applying shingles. Here's a shot of the lower roof section, almost done.

29 Find the angle for the trim. The intersection of the lower roof panel (J) and the clerestory wall (K) is a potential trouble spot for leaks. I made a piece of wooden trim (O) to cover this area, which makes for a nicer look and helps with the waterproofing. To fit properly, the bottom edge of the trim strip needs cut at an angle, and I determined the angle by using a simple angle finder.

30 Cut the angle on the trim. By angling the blade on my table saw, it was easy to create a piece of trim (O) that fit perfectly. I also decided to cut the same angle on the top of the trim strip, so that it would shed water better (not shown).

31

32

33

31 Install the trim. After being painted, the trim strip (O) creates a nice clean transition between the roof panel and the wall above it. Nail it in place. Install doors (M, V).

32 Add some decoration. The space below the gable seemed like the logical spot for some ornamentation. I didn't want to overdo it, but a large white star did the trick.

33 Build the base. When I arrived on site, I built a simple but sturdy base to hold the coop up off the ground. Screw together some sturdy 4x4 legs (P) and some ¾" (17mm) plywood stretchers (Q, R). This coop is quite roomy on the interior, so nest boxes (S, T) and roosts (U) of nearly any style can fit inside.

Tea Egg, Chinese: Soak cracked, hard-boiled eggs in black tea and Chinese five-spice.

ANDREW STONE
Salt Lake City, Utah

Andrew Stone is a realtor by trade; more specifically, he's an eco-broker. I hadn't heard of this before, but it turns out to be pretty much what it sounds like: he specializes in helping his clients save money and live comfortably through energy efficiency, renewable energy, and environmentally conscious choices. When I heard that, it all came together—of course he's the kind of guy who would be into backyard chickens! They're a perfect expression of his commitment to local, grassroots food production, and when Salt Lake City relaxed its ordinance and allowed the keeping of domestic fowl, Andrew jumped on the opportunity.

At Andrew's childhood home in California, his family raised lots of animals as he was growing up. He fondly recalls having goats, horses, ducks, quail, geese, pheasants, and of course, chickens. These early experiences provided him with a great springboard when he decided to get back into the poultry game as an adult. "I have kids now," he explained, "and it is nice for them to see where their food comes from."

organic gardening

Andrew is a prolific organic gardener, and the leap to having chickens again brought some pretty appealing benefits. He's been growing much of his family's food for a long time, and he's understandably proud of the fact that his chickens are now providing him with terrific fertilizer as well as fresh eggs on a daily basis. Chicken manure is the animal manure richest in nitrogen, phosphoric acid, and potash—the three chemicals that commercial fertilizers are measured by. I can attest to the value of this myself—our "chicken-enhanced" soil is a real miracle worker.

⬆ Andrew utilizes the chicken droppings in his tumble composter, seen inside the run here. His garden, chock full of leeks, brussels sprouts, butternut squash, kale, chard, carrots, kohlrabi, and much more, flourishes with the addition of the natural fertilizer.

 Andrew's chicken setup is a great example I think is worth emulating. His chickens have plenty of room to roam around, and they can choose between spending their time in sunny or shady areas, depending on the weather. Their run is roofed in and is easily secured from predators when the chickens are put in for the night.

If the chickens are fighting over food or water, add another feeding station.

The chickens sleep and lay their eggs inside a cozy, weathertight coop situated within the run. The sunrise motif, chicken prints, and Victorian spindle silhouette details add quite a bit of personality to the coop. Also, notice Andrew's chicken ladder—it is constructed differently than the way I've shown in this book, which just goes to show you that there is a lot of room for creativity when building coops!

The roof on this run makes me feel jealous of Andrew's setup—it is framed with 2x4s and covered with corrugated panels. Some of the panels are clear and let sunlight in, while others are semi-opaque and block most of the sun in the summer.

Both the feeder and water container are suspended off the ground, which is great for two reasons: it prevents the chickens from making much of a mess, and it also renders the food inaccessible to rodents, which is always a good thing.

meet the chickens

Andrew has six chickens, which he raised from day-old chicks. He has a variety of breeds, including Buff Orpingtons and Ameraucanas, and they're laying four eggs a day at this point. When he was planning out their future headquarters, he identified a perfect spot next to the garage that was already bordered on two sides by a fence. This cut down some of the time and effort required to build a run— two sides were already built. With the addition of a roof, the birds have plenty of shade to help them beat the heat in the summer. The chickens have a small coop inside the run where they lay eggs and sleep at night.

During the day, they love to get out and peck around for food—which there is plenty of, thanks to the garden's bounty. Andrew actually placed the compost pile within reach of the chickens, who perform free labor every day as they dig through it and speed the process of building new soil for next year's garden beds. In fact, Andrew remarked with a smile, it turned out that the chickens initially loved the garden a little too much—he began by letting the chickens roam free, but very quickly realized that they ate a lot more than he would've imagined. They were particularly partial to rhubarb, and they had a troublesome penchant for digging up freshly planted seeds, but once he erected a wall to limit their access, these headaches disappeared.

challenges

I asked Andrew what his biggest challenges had been, and it was funny to note that he couldn't really think of any. "They're pretty easy," he summarized, "once you're up and running." When pressed, he did relate the story of Lucy, the chicken who began crowing one day and was then renamed Luke. He easily found a new home for Luke on a farm that was looking for a rooster to accompany its flock of hens, so that worked out fine, and Andrew found a replacement hen on another farm about a half hour away. At the time of my visit, the new hen had only been on the premises for a day or so, and Andrew mentioned that the flock was undergoing the usual adjustment issues. "She's definitely at the bottom of the pecking order," he said, but she has plenty of places to hide during the week or so that it would take for the rest of the flock to accept her. While people who are new to keeping chickens may have a harder time with a situation like this, Andrew's a veteran and he wasn't worried. Neither was I—I could see that Andrew's spacious layout would allow plenty of room for everybody.

⬆ Andrew's flock has six chickens; he prefers the Ameraucana and Buff Orpington breeds. Ameraucana lay fun colored eggs, while Buff Orpingtons are large, sturdy chickens that handle winter weather pretty well.

tips

Andrew had some great tips for people who are thinking of raising chickens of their own:

- **Follow the letter of the law.** Some people fly under the radar, but Andrew is an advocate for playing by the rules. "This way, when you have a permit, you're covered if you ever have neighbors who have a problem. You can prove that you're doing everything right."

- **Don't bother with roosters inside city limits.** "It's not worth it," he said, and I agree. For one thing, it is illegal in many cities, and complaints from the neighbors are inevitable.

- **Go for it!** Andrew is a great advocate for backyard poultry, and he insists that just about anybody can do it. "There's really nothing to it," he says, and the benefits are huge!

Chickens will eat most fruit, vegetable, or bread scraps.

GYPSY HEN CARAVAN
A Whimsical Nomadic Coop

Partway through the writing of this book, I became fascinated with the art, craftsmanship, and history of gypsy vans. It wasn't something I'd had a lot of exposure to in the past, but after seeing a couple of great photos in a book, I couldn't resist some web-surfing that brought up a treasure trove of mind-blowing pictures. A couple of weeks later, I was discussing the concept of "chicken rickshaws" with a friend and voila!—the idea was born to build a coop that had the look and feel of a vintage gypsy wagon. The concept is actually quite practical, too: I have a number of clients who house their flock in different locations, depending on the season, so having a mobile coop is pretty important for them.

In terms of scale, I had to make some adjustments to create a coop of a practical size—original gypsy wagons could easily by as much as 10' (3000mm) tall, and the wheels alone often measured 5' (1500mm) high. So, I did take some liberties in modifying the proportions to come up with a finished product that would fit comfortably into its setting—in this case, a small city lot that would've been dwarfed by a full-size wagon. As for the wheels, I had originally considered using bicycle wheels, but decided that they wouldn't look quite right, so I made some wheels from ¾" (17mm) plywood that are a better match.

This project is not a replica by any means—a person could spend a year or more trying to build a perfect reproduction—but I think it captures some of the whimsical spirit of the originals that inspired it. And, as a fun historical aside, I've seen some wagons that actually had chicken cages built into them. So maybe this idea for a coop isn't so far-fetched after all!

The chicken coops are clearly seen on the back underside of this gypsy wagon.

If you're interested in reading more about gypsy wagons, I highly recommend the following website: *www.gypsyvans.com*.

Over easy = An egg lightly fried on both sides, but with a runny yolk and some runny whites.

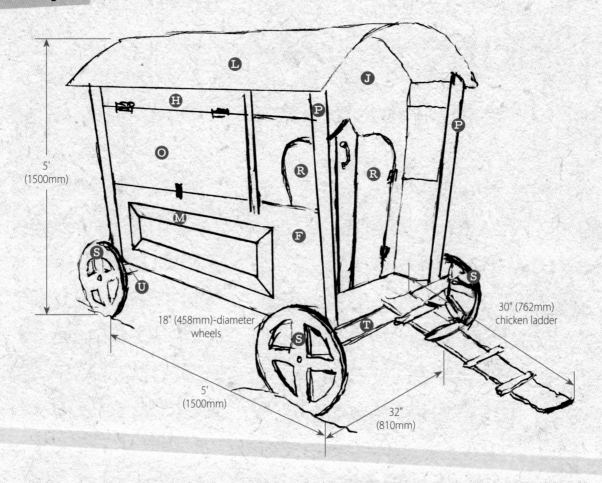

5'
(1500mm)

18" (458mm)-diameter
wheels

30" (762mm)
chicken ladder

5'
(1500mm)

32"
(810mm)

1 Begin building the floor. I
decided to build the coop first and the
wheeled base second. I reinforced the
long edges of the coop floor (A) with
a pair of 2" x ¾" (50 x 20mm) slats (B).
These slats were beveled at a 7° angle
on the bottom edge so that the sides
of the coop would flare out.

2 Attach the long side supporting
slats. To secure the slats (B) to the
floor, I put in screws every 6–8"
(152–205mm).

MATERIALS LIST

item	material	dimensions	quantity
Ⓐ Platform (floor)	¼" (6mm) plywood	5' x 2½' (1500 x 800mm)	1
Ⓑ Long side reinforcing floor slats	¾" (19mm) pine	5' x 2" (1500 x 50mm)	2
Ⓒ Short side reinforcing floor slats	¾" (19mm) pine	2½' x 2" (800 x 50mm)	2
Ⓓ Back wall panel	½" (11mm) plywood	4' x 4' (1200 x 1200mm)	1
Ⓔ Back panel reinforcing side slats	¾" (19mm) pine	3' x 2" (900 x 50mm)	2
Ⓕ Side wall panels	½" (11mm) plywood	5' x 1½' (1500 x 500mm)	2
Ⓖ Vertical corner slats	¾" (19mm) pine	3' x 3" (900 x 75mm)	2
Ⓗ Top rails	¾" (19mm) pine	5' x ½' (1500 x 200mm)	2
Ⓘ Filler slats	¾" (19mm) pine	12" x 5" (305 x 128mm)	4
Ⓙ Front gable panel	½" (11mm) plywood	4' x 2½' (1200 x 800mm)	1
Ⓚ Roof nailers	2x10 (38 x 235mm)	8' (2400mm) total	1
Ⓛ Roof	⅛" (3mm) plywood	5½' x 4' (1700 x 1200mm)	1
Ⓜ Accent panel trim	Paint-grade trim	8' (2400mm)	6
Ⓝ Stars	¼" (6mm) plywood	10" (255mm) squares	3
Ⓞ Window flaps	¾" (17mm) plywood	40" x 12" (1015 x 305mm)	2
Ⓟ Corner posts	¾" (17mm) fluted molding	39" (990mm)	2
Ⓠ Front door trimmer	¾" (17mm) plywood	39" x 4" (990 x 102mm)	2
Ⓡ Front doors	¾" (17mm) plywood	32" x 15" (815 x 380mm)	2
Ⓢ Wheel blanks	¾" (17mm) plywood	19" x 19" (485 x 485mm)	4
Ⓣ Axle supports	2x4 (38 x 89mm)	8' (2400mm) total	1
Ⓤ Axles	⅝" (16mm) threaded rod	4' (1200mm)	2
Ⓥ Rear doors	¾" (17mm) plywood	1½' x 1½' (500 x 500mm)	2
Ⓦ Nest box sides	¾" (17mm) plywood	1' x 1' (300 x 300mm)	3
Ⓧ Nest box top	¾" (17mm) plywood	28" x 12" (710 x 305mm)	1
Ⓨ Roosting bars	¾" (17mm) pine	3' x 1" (900 x 25mm)	2
Screws			
Hinges			
Paint			

The number of folds in a chef's hat = How many ways the chef can cook eggs.

3 Create short side supporting slats. To stiffen up the back edge of the floor, I used a slat (C) similar to the ones used on the long edges. However, this slat didn't need beveled on the bottom, and it did need to have its ends cut at a 7° angle so it would fit properly. To mark the ends, I tacked the slat in place and drew a line at the intersection of the long and short slats.

3

4 Begin the back panel. Because the back panel (D) lines up with the side panels (F) along its vertical edges, it needs to flare out at the same 7° angle as the bottom slats (B). To lay out the correct shape, find the center of the blank (2' [600mm] across the 4' [1200mm]-wide panel), and from there, measure 15" (380mm) to the left and right of the center point and mark the 30" (762mm) width of the floor.

5 Begin outlining the back panel. I set my angle finder to 7° and aligned it with the marks delineating the outside edge of the back panel (D). Drawing a line here established the beginning of the outline for the back panel.

6 Extend the lines. Use a straightedge to extend the angled line. The finished height of the side will be 34" (865mm), so that's how long the line needs to be. Repeat on the other side of the blank.

4

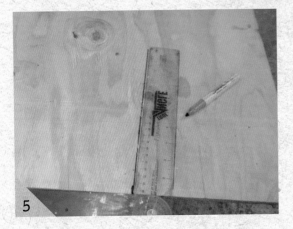

5

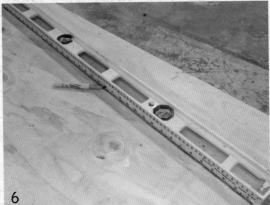

6

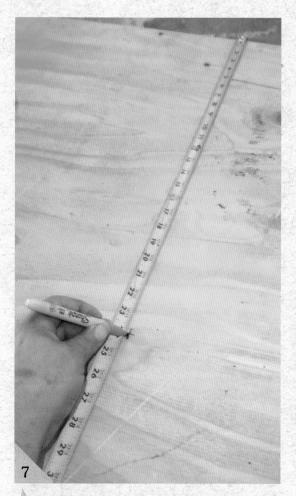

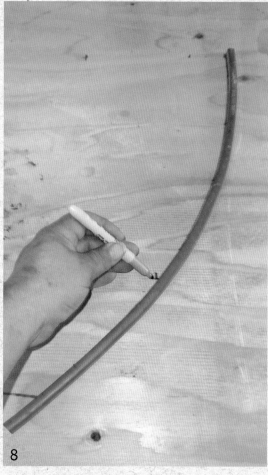

7

8

PREDATOR-PROOFING

These folks fortified the bottom edge of their chicken run with loose stones from the property, and they also installed an additional layer of ½" (12mm) wire mesh to create an extra barrier.

7 Begin defining the arched top. After I laid out the position and length of the sides, I began to think about the arched top. A flat top would have made things easy, but that wasn't the goal here. Once again, I needed to find the center, this time at the top of the back panel (D), so that I could think about the rise and run of the arch.

8 Mark the curve. I decided that a rise of about 7" (178mm) looked good, so I marked 7" (178mm) above the center mark and laid out a curve using a flexible curve. There's no need to make this complicated—just come up with a curve that looks nice. You'll notice that I'm only thinking about one side of the arch to begin with—that's because I planned to use the cutout as a pattern for the other side.

 World Egg Day is the second Friday in October.

9 Cut out the first half of the curve. Using a jigsaw, I cut along the top of the arch and then cut away the scrap. The second cut doesn't have to be pretty.

10 Mark the second half of the curve. By lining up the cut-off with the center mark and the top of the left-hand side of the back panel, it was easy to complete the arch.

11 Examine the arch. This approach works great, but it is worth taking your time on the first half, because any mistakes or discrepancies you make there will be transferred to the other side. You can see that I had a small hiccup in the middle of each half of the arch that I had to sand out.

9

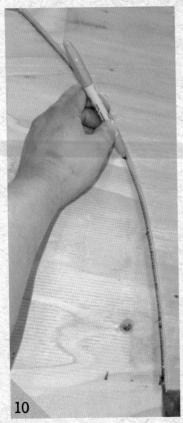

10

11

12

13

14

12 Attach the back panel to the coop. After the shape of the back panel (D) is done, it can be attached to the slat at the rear of the coop.

13 Attach back side slats. To stiffen up the back panel (D), and to provide a place to secure the side panels (F), I attached a pair of slats (E) along the edge of the sides with screws and glue.

14 Install the bottom side panels. Because I wanted large windows on the sides of the coop, I didn't start out using a blank that covered the whole side. It is easier to rip sections of ½" (11mm) plywood to cover from the floor to where I wanted the bottoms of the windows to be located—about 36" (915mm) (F). However, if you want to have window coverings (O) like I eventually decided to, you might want to construct the sides out of large pieces of plywood and plunge-cut the window openings.

For easy whipping, use room-temperature egg whites and a dry, clean bowl and whisk.

15

15 Add vertical slats. To frame around the window openings and provide a front corner post of sorts, I screwed a vertical slat (G) to the side panels (F). I held back the vertical slat about 10" (255mm) from the front of the coop to create a porch—this is a common feature on many gypsy wagons.

16 Clamp the top strips. To fill in the area above the windows and below the roof, I ripped a piece of ¾" (19mm) stock to 6" (152mm) wide and clamped it into place (H). Attach with screws.

17 Insert filler slats. To create a flush side, I inserted 5" (128mm)-wide slats (I) on top of the vertical pieces to the left and right of the windows. This made for a neater look, and it would also make it easier to apply trim later on.

16

17

18 Create the shape of the front gable. Attach the front panel (J) in place. I wanted the doors at the front of the coop to be pretty flamboyant, so I came up with an arabesque shape for the top of the door opening. Although this contour is more sophisticated than the curve at the top edge of the back panel, the actual technique was the same—I laid out and cut out one half of the arch and then used it as a template for the other side. In this case, I was working on a vertical panel, so I used blue painter's tape to hold the flexible curve in place.

19 Trace the first half of the front gable. With the flexible curve removed, you can see the pencil line delineating the shape that I had in mind.

20 Trace the second half of the front gable. Again, blue painter's tape came to the rescue—this was easier than trying to hold the pattern steady while I traced.

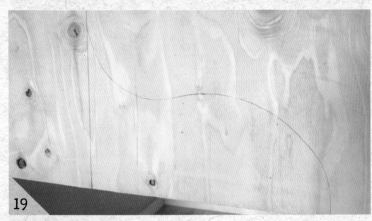

 There are more than 150 varieties of domestic chicken.

21

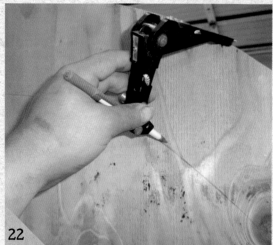

22

21 Complete the gable cutout. The completed cutout provided exactly the kind of flourish that I was hoping for.

22 Draw the back window opening. As the design evolved, I decided to put a window opening near the top of the back panel (D). I wanted the top of the opening to follow the shape of the arch above it, so I used a compass set to 4½" (115mm) wide to scribe a line below the top of the arch.

23 Cut out the back window. Here's the completed window opening—the sides were angled to match the sides of the coop, and the bottom was flat. Use a jigsaw to cut out the shape. If you want to have doors over this window, plunge-cut the opening and save the pieces for Step 46.

23

24

25

26

24 Place nailers. To support the barrel-shaped roof, I decided to attach a 2x10 nailer (K) at the front and back of the coop. To make sure it matched the shape of the arch, I clamped it in place temporarily and then traced the arch panel.

25 Attach nailers. Once the nailers (K) had been cut to shape, I put them in place permanently using a nail gun.

26 Begin attaching the roof. The roof of the coop (L) is made from a sheet of ⅛" (3mm)-thick Maranti plywood. The exact material isn't critical—it just has to be thin so it will bend easily. As luck would have it, I didn't have to rip this sheet down—its 4' (1200mm) width provided just the right amount of overhang on the sides of the coop—but I did need to crosscut it to 66" (1675mm). Then I put it on top of the coop and centered it so that the overhangs were equal all around. Next, I put a couple screws down through the roof into the nailer (K) at the highest point of the roof (L).

The longest recorded chicken flight was 13 seconds.

27 Screw down the roof. After tacking the roof (L) in place, it's easy to screw it to the nailer (K) atop the front and back panels, and to the top edges of the side panels (H). Ensure the edge of the roof is flat—any spots that don't flatten out can be fixed with a screw or two.

28 Locate the nest box access panel. I figured the back of the coop was a logical spot to put a nest box access panel. I laid it out with a framing square and a straightedge.

29 Cut out the nest box access panel. I used a jigsaw to carefully cut out the panel, making sure to keep it in good shape because I planned to use it for the door. Plunge-cutting with the jigsaw made this a snap. Attach the panel with hinges on the bottom.

30 Paint the body. It's time to paint. This may seem counter-intuitive, but it is actually a lot easier to lay down the body color now and pre-paint all the trim and secondary parts. This saves a lot of time and hassle in masking things off and trying to cut in neatly.

27

28

29

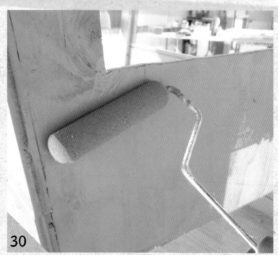

30

31 Paint the trim. This kind of planning ahead makes life a lot easier—I lined up all of my trim stock (M, P) and spray painted it with a gold metallic paint, which imitates the gold-leafing found on many original wagons. It only took a couple of minutes.

32 Cut out and paint the stars. I decided on a star motif for the area above the front door. Draw the stars freehand to a size that looks right, then cut them out on a band saw. I painted the stars (N) prior to attaching them, which made for a better result and an easier process.

33 Add side panel embellishment. The large open space on the lower parts of the side panels (F) seemed like a natural place for some embellishment, so I laid out a rectangle to paint another color. I didn't worry about getting a perfectly neat edge along the sides of the rectangle; I'll overlap the trim (M) about ¾" (20mm) to create a clean transition.

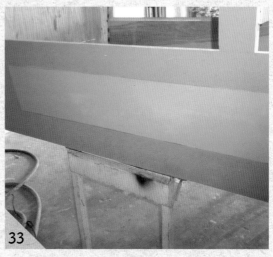

Avgolemono, Greek: A soup with egg, lemon, broth, rice or orzo, and chicken.

34 Attach trim around the side accent panels. Attach the trim (M) with nails, then fill the holes with wood putty and paint over them. The gold paint provided just the sort of regal feel I had in mind. The area inside the frame can be dressed up in any number of ways, or just left alone.

35 Attach the window flaps. I debated how best to treat the window areas—in the end I decided to hang a ¾" (17mm)-thick panel (O) from hinges located at the top of the opening. Because the chickens will only hang out in the coop when they're laying or sleeping, the windows will be closed most of the time, but I wanted the flaps to be operable in case light and ventilation were deemed useful.

36 Attach flap closers. To keep the flaps closed, I made a simple latch from a scrap, painted it gold, and screwed it loosely in place so it could be swiveled.

34

35

36

37 Add front corner posts. To dress up the front of the coop, I used fluted molding (P) at the front edges of the sides. This created a corner post to visually define the porch a bit better. I also added gold trim on the back of the coop to frame around the opening that provides access to the nest boxes.

38 Cut out the door trimmers. No coop is complete without a set of doors to keep its residents safe and secure. Mounting the doors required the installation of an angled trimmer (Q) along each side of the door opening. I had to notch the trimmer at the bottom so it would fit neatly. I cut the tops and bottoms at a 7° angle.

PRE-COOP LIVING SITUATION

Baby chicks don't need anything fancy for their first residence. We used an old cabinet that was laying around the shop, but most people do just fine with a cardboard box. You'll want to make sure that you have something to place over the top to keep them inside, and plenty of dry bedding for them to snuggle in.

There are 7 types of comb: pea, cushion, buttercup, single, rose, V-shaped, and strawberry.

39 Paint and install the trimmers.
When the trimmers (Q) fit, I painted them. After the paint had dried, I secured them with several screws.

40 Mark and cut out the front doors. Hold the rectangular blank (R) for the first door behind the front panel and trace on the profile of the opening. To mark the centerline, use a straightedge to draw a vertical line from the top of the door on down. Cut out the door and discard the scrap. Repeat for the second door.

41 Install the hinges and doors.
Attaching the door (R) is easier when you hang the hinges on the trimmer first. After that, you can hold the door in place and sink screws into it, adjusting the fit up and down or side to side as needed.

39

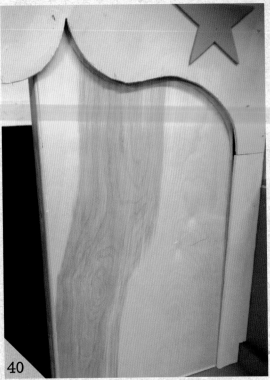

40

41

42

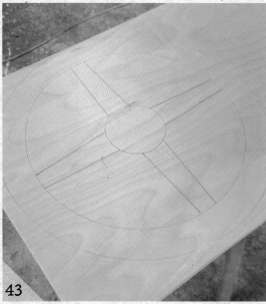

43

44

42 Paint the front doors. I decided to add another trim element to the center of the doors. It is another way to add color, shape, and variety to the whole composition, but it also serves a practical purpose. If your doors didn't meet up evenly in the middle, this hides the discrepancy—and nobody will know but you.

43 Lay out the wheels. I made the wheels (S) from ¾" (17mm) plywood. I began by drawing one with a compass and straightedge. The diameter of these wheels is 18" (458mm), with a 1¾" (45mm)-thick rim.

44 Cut out the wheels. Cut out the wheel (S) with a jigsaw. It can now serve as a template for the other wheels (S). Cut out all the wheels and paint them.

 Hens can live up to 20 years.

45 Install axles. To attach the wheels (S) to the coop, I ran a 42" (1068mm) length of 2x4 (T) beneath the coop at the front and back edges. I screwed a set of four 4" (102mm)-long blocks (T) below the ends of both 2x4s (T) to support the axle (U), which I made from a 4' (1200mm) length of threaded rod. The blocks (T) also provide a large, flat surface to support the wheels (S) and keep them from wobbling. Drill a ¾" (20mm) hole in the middle of the 2x4 blocks (T).

46 Make the back doors. As a final detail, I made a set of doors to fit into the opening on the back of the coop. I held the oversized door blank (V) inside of the opening and traced the profile onto its surface. Once the door was cut out on the band saw, it fit neatly into place. If you decided in Step 23 that you wanted doors, just use the pieces you cut from the back panel at that point.

45

46

47

47 Attach the back doors. Paint the doors and attach them with hinges. The only remaining detail is to make a latch to keep the birds secure at night, as was done in Step 36.

EGG IN A BASKET WITH BACON

Photo courtesy of Scott D. Feldstein.

Bread, sliced
Egg
Bacon, several slices

Use a small glass to remove a circle of bread from the center of each bread slice. Fry up the bacon in a pan over medium-high heat. When the bacon is done, arrange the slices into a row. Place the bread slice over the bacon. Crack the egg into the hole; after egg has cooked in place, flip the bread. Cook to desired doneness.

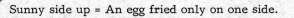

Sunny side up = An egg fried only on one side.

HOW THE CHICKEN CROSSED THE ROAD
A Movable Coop that Makes Cleanup Easy

The inspiration behind this coop was functionality—what kinds of features could be added to make the most practical, easy-to-use coop possible? I worked with a great client to brainstorm on this theme, and we came up with a couple of priorities:

- The coop needed to be mobile, and

- We wanted to streamline the cleanup process.

Adding wheels to one end of the coop worked out as a way of providing mobility. One or two people can easily wheel it around, and this type of structure fits into the category of "chicken rickshaws," which are a nifty coop subgenre I've heard about but never seen. Figuring out how best to install and support the wheels took some head-scratching, but I came up with a simple solution that works great.

To simplify the cleanup process, we came up with a couple of ideas. I put a large set of doors on the end walls of the coop to provide great access to the coop's spacious interior, and this paved the way for what I think is a pretty novel concept: the poop tray. My client mentioned this to me, and it made a lot of sense. If we situated the roosts directly above a pair of removable trays, the trays would collect the majority of the chicken droppings, which could then be disposed of quite easily and with minimal fuss. We reasoned that the trays could slide in a set of tracks so they wouldn't get moved around, and the roosts could simply be moved out for hosing off. We also hoped to prompt the birds to use the roosts by modifying the tops of the nest boxes—they're popular places to perch most of the time, but the angled design here prevents this.

Cochin, Silkie, and Phoenix are beautiful ornamental breeds.

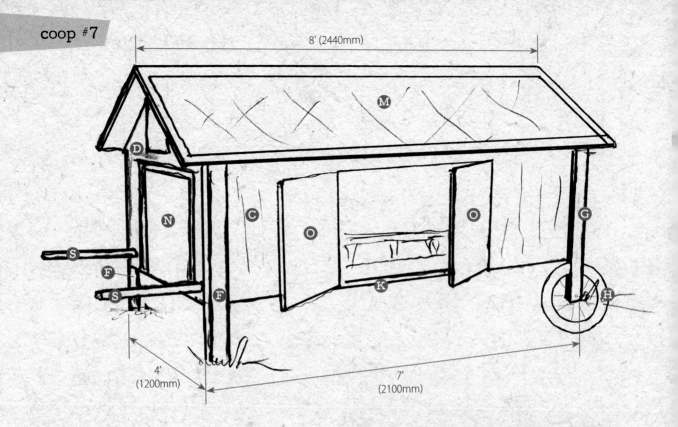

8' (2440mm)

4' (1200mm)

7' (2100mm)

1 Construct the bottom frame. I began this coop by constructing a frame from 2x4s (A, B). To join the parts, I used galvanized angle brackets. I like Rigid-Tie Angle by Simpson in particular because it features a flange that allows screw installation on the inside frame face, which adds quite a bit of stiffness.

2 Attach the side walls. I built the sides of the coop (C) out of ⅜" (8mm) plywood ripped down to 2½' (800mm) wide. With the frame sitting directly on the floor, it is a straightforward matter to screw the sides (C) to the frame. I suggest spacing the screws about 1' (300mm) apart.

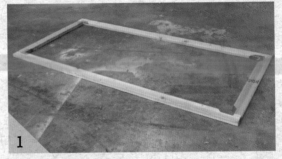

1

2

MATERIALS LIST

item	material	dimensions	quantity
Ⓐ Long side floor framing	2x4 (38 x 89mm)	7' (2100mm)	2
Ⓑ Short side floor framing	2x4 (38 x 89mm)	41" (1040mm)	2
Ⓒ Side wall panels	⅜" (8mm) plywood	7' x 2½' (2100 x 800mm)	2
Ⓓ End wall panels	⅜" (8mm) plywood	3½' x 4' (1100 x 1200mm)	2
Ⓔ Interior corner bracing	2x4 (38 x 89mm)	28" (710mm)	4
Ⓕ Front legs	2x4 (38 x 89mm)	51" (1295mm)	2
Ⓖ Rear legs	2x4 (38 x 89mm)	45" (1145mm)	2
Ⓗ Wheels	Bicycle rims with wheels	16" (405mm)-diameter	2
Ⓘ Axles	½" (13mm) bolts with lock nuts	7" (178mm)	2
Ⓙ Window trim	Scrap cedar 1x2 (19 x 38mm)	14" (355mm)	8
Ⓚ Platform (floor)	½" (11mm) plywood	7' x 4' (2100 x 1200mm)	1
Ⓛ Rafters/roof cleats	2x4 (38 x 89mm)	7' (2100mm)	3
Ⓜ Roof panels	¾" (17mm) plywood	35" x 30" (890 x 762mm)	6
Ⓝ End panel flap doors	¾" (17mm) plywood	14" x 20" (355 x 508mm)	2
Ⓞ Main side doors	¾" (17mm) plywood	26" x 22½" (660 x 572mm)	2
Ⓟ Tracks for poop tray	2x4 (38 x 89mm)	82" (2085mm)	2
Ⓠ Poop trays	⅜" (8mm) plywood	40" x 18" (1016 x 458mm)	2
Ⓡ Wheel supports	2x4 (38 x 89mm)	1' (300mm)	2
Ⓢ Handles	2x2 (38 x 38mm)	40" (1015mm)	2
Ⓣ Roll roofing		48sf (4.5sm)	
Ⓤ Felt paper		48sf (4.5sm)	
Ⓥ Miscellaneous strips for trim	¾" (17mm) pine or similar	15' (4600mm) total	
Ⓦ Nest box sides	¾" (17mm) plywood	15" x 12" (380 x 305mm)	5
Ⓧ Nest box top	¾" (17mm) plywood	52" x 13" (1320 x 330mm)	1
Ⓨ Roosting bars	¾" (19mm) pine	78" x 1" (1980 x 25mm)	2
Ⓩ Roost supports	2x4 (38 x 89mm)	16" (405mm)	3
⒜⒜ End panel window covers	¾" (17mm) plywood	11" x 11" (280 x 280mm)	2
Galvanized angle brackets			
Screws			
Hinges			
Staples			
Nails			
Chicken wire			
Brass handle			
Latches			
Roofing adhesive			

 There are more chickens in the world than people.

3 Prepare to add the end walls.
Without the end wall panels (D) in place to support the sides (C), the whole structure isn't particularly strong, and the walls (C) may lean in a bit. This is not a cause for concern—adding the end walls (D) will firm things right up.

4 Lay out the end walls.
The end walls (D) are ⅜" (8mm) plywood. I laid out a roofline as follows: the height of the panel is 3½' (1100mm) and the start of the pitch is at 2½' (800mm), meaning the taper rises 1' (300mm) over a 2' (600mm)-wide span, sometimes described as a 6:12 pitch. I used a framing square to lay out a nice wide door (30" x 27" [762 x 685mm]) to make cleanout a snap, and added a small 11" x 11" (280 x 280mm) window below the gable for some extra ventilation.

5 Cut out the end wall doors and windows.
When cutting out the doors, it is important to start and stop the cuts cleanly in the corners so that the "waste" pieces can be re-used as doors. I used a circular saw for this operation, but a jigsaw would work well, too. The windows aren't critical—they are such small pieces, it is easy to find scraps to fit.

3

4

5

6 Attach the end walls to the coop. Screw the end walls (D) through the base frame (A, B) and reinforce the corner where the wall panels converge by adding interior corner bracing (E). I used 2x4s with screws and glue, and I made sure to cut the 2x4 shy of the top of the walls so that I would have space to tuck a 2x4 cleat (L) in there—more on this later.

7 Attach the legs. With the shell of the coop assembled, I turned my attention to making and adding the legs (F, G) that hold the coop up off the ground. These are just 2x4s, mitered at the top to match the roof angle (26½°) so the roof sheathing can overhang a few inches. I started with legs all the same length, but you can use lengths marked on the materials list to avoid cutting in the next step.

8 Locate the wheels. Because I planned to put a set of wheels (H) on the far end of the coop, I had to shorten the legs on that end (G) and drill a hole for the axles (I)—if you used 45" (1145mm) legs for the back, you won't need to cut. To lay this out, I lined up the wheel (H) with the bottom of the rear legs (G) and marked for the axle (I). The leg could then be cut a few inches up from the bottom.

Dual-purpose chicken breeds can be raised for meat and eggs.

9 Drill the axle holes. I had planned on using a ½" (13mm)-diameter axle, so I drilled a %16" (14mm) hole to allow just a little bit of wiggle room.

10 Review your progress. This photo shows how the legs are different lengths—prior to installing the wheels, the whole thing actually sloped downhill a little bit.

11 Mount the axle and wheels. My first inclination was to use an axle that ran from one side of the coop to the other, but the axle sagged more than I was comfortable with. Still, it was a start, and it got the coop leveled out so I could continue working while I brainstormed a solution. However, you may wish to install the ½" (13mm)-diameter bolts (l) now.

12 Cut out the nest box access. The long back side of the coop is reserved for nest box access. Cut out a large 4' x 1' (1200 x 300mm) rectangle—you will be using it as a door, so cut carefully.

9

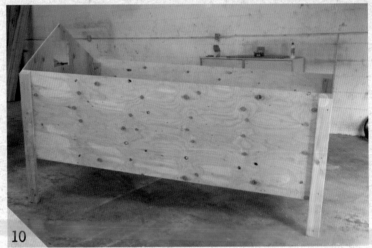

10

11

12

13 Hinge the nest box access door. After I installed three hinges, the door worked out great: it allowed for access to all the nest boxes at once, and putting the hinges at the bottom meant that a person wouldn't need to use one hand to keep it open.

14 Cut out the side door. Across from the nest box area, I cut out a large 44" x 24" (1118 x 610mm) opening for day-in-and-day-out access.

15 Trim the end wall windows. To trim out the small window on the end walls (D), I used ¾" (19mm)-thick scraps of cedar (J). I assembled the trim as a unit before mounting to make sure the miters were all tight. When the glue dried, I attached the frames to the outside of the openings and secured them with screws from inside the coop. The opening of the frame is about 1" (25mm) smaller than the window—this hides the plywood edge and creates a more finished look.

16 Install the bottom. I used ½" (11mm)-thick plywood for the bottom (K). The only trick was that I had to notch the corners with a jigsaw so the bottom could fit properly around the corner supports (E). This only took a few minutes, and made for a nice result.

Tamagoyaki, Japanese: A slightly sweet omelet with soy sauce and sugar, rolled to serve.

17

17 Trim the end panel doorways. Trim out the door openings. Set the trim (V) in about ½" (12mm) to conceal the edge of the plywood.

18 Build the nest boxes. This coop was built for about fifteen chickens, so four nest boxes was about right. A good rule of thumb is to provide one nest box for every four birds. It was easier and faster to build one large unit and just add a few dividers (W) than to build separate boxes. Cut a wedge off each divider (W) so the top edge slopes from 15" (380mm) in the back to 12" (305mm) in the front. I put a small lip—the exact height is not critical—at the bottom of the boxes to keep the bedding material from spilling out.

18

19

20

21

19 Place the nest boxes. The nest boxes fit perfectly into place and they should work out great. They're not fastened so they can be removed for cleaning, but hopefully the sloped roof design will prevent perching and the subsequent droppings. This will make for less maintenance down the road.

20 Install the roof cleats. As I mentioned earlier, I added a 2x4 cleat (L) at the top of both side walls (C). This provides a place to secure the lower edge of the roof panels (M). I also installed a 2x4 (L) at the peak of the roof as a way to support the top edge of the roof panels. Angle the 2x4s so the roof panels (M) will be flat.

21 Attach the roof panels. The total length of this coop was 96" (2400mm) (84" [2100mm] for the coop, 6" [152mm] of overhang on each end), so I couldn't use just one piece of plywood as roof sheathing. This wasn't a problem—I just cut separate panels (M) and installed them side-by-side to the underlying 2x4s (L).

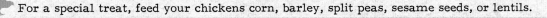

For a special treat, feed your chickens corn, barley, split peas, sesame seeds, or lentils.

22

23

24

22 Staple on the felt paper. As insurance against future leaks, I added a layer of felt paper (U) to the roof and secured it with staples. My trick is to put on pieces of paper slightly larger than the roof and then trim them to size with a utility knife. This is easier than cutting the felt paper exactly to size and then trying to install it precisely—it can be hard to align when you're working alone. However, if you're going to use adhesive in Step 32 as I do, you should wait to trim the felt paper until then.

23 Attach the end panel doors. The doors (N) on the end of the coop have hinges at the bottom. It is faster and easier to install one large door than two smaller ones, and it saves a couple of bucks by requiring half as many hinges. Use the wood you cut out in Step 5 if it looks OK. Install a latch on the top of the doors.

24 Trim out the end panel doors. I trimmed out the edges of the doors (N), but I had to miter the bottom edges of the vertical trim (V) so that the door could swing down all the way. A blunt 90° cut would've gotten in the way.

25

26

27

25 Install wire on the windows. I stapled chicken wire on the inside of the small ventilation windows. This will keep chickens in and predators out, and the extra airflow will be a real plus.

26 Install main side doors. The main side doors (O) were hinged at the sides, and I ran a strip of wood (V) up the center to flange over the gap between the two doors. A small brass handle is a nice touch here.

27 Create the poop tray tracks. I've never seen a poop tray built into any other coop, but it is a really practical idea that saves time during cleanout. The roosts are above a pair of removable trays, so the droppings are consolidated in one place. The trays slide in a simple set of L-shaped tracks (P) made on the tablesaw. Two cuts on adjacent faces of a 2x4 removed the waste material, and I was done in just a few minutes. Note that this operation will require you to change the blade height and fence position between cuts. The exact location and size of the cut-out isn't critical.

JUST ADD STRAW

Although it isn't necessary, these chickens do seem to love having a layer of straw on the floor of their run. They spend a lot of time digging through it with their feet and hunting for bugs.

Over medium = A lightly fried egg with runny yolks.

28

29

28 Install the tracks. Install the tracks (P) parallel to each other about 18½"–19" (470–485mm) apart, using a series of countersunk screws.

29 Create the poop trays. The trays (Q) themselves are 18" (458mm) wide by 40" (1015mm) long. I considered putting a handle on the end, but figured that a simple hole would function just as well. I used a 1¼" (32mm)-diameter spade, or paddle, bit to make the holes.

30 Make the roosts. The roosts (Y, Z) are lightweight and removable, and they fit neatly atop the trays. You could build them in two sections, but I didn't see a downside to building a single long one. Screw the long dowels (Y) into the roost supports (Z).

30

31 Install wheel supports and permanent axles. As I worked out a good way to support the wheel, I considered welding up something akin to a bicycle fork, following the logic that the wheel needed to be supported on both sides. I then realized that adding a 2x4 (R) would do the trick just fine, and 7" (178mm)-long bolts with lock nuts (I) held everything in place quite well.

31

32

32 Apply roofing adhesive. When it came time to install the roofing, I cut back the felt paper (U) a few inches along the edges because I wanted to put in a layer of adhesive to secure the roll roofing to the sheathing (most home improvement stores carry an inexpensive product for just this purpose).

33 Apply the roofing. Rolling out the roofing (T) by yourself is tricky, but if you keep it rolled up and then secure one end with staples, you should be able to manage it. When in doubt, get help.

EGGS BENEDICT

Photo courtesy of Paul Goyette.

English muffin
Canadian bacon
Egg
Hollandaise sauce (see page 84)

Poach the egg; serve it on an English muffin topped with a slice of Canadian bacon, and add a dollop of Hollandaise sauce over everything.

VARIATIONS:
Country Benedict: Substitute a biscuit, sausage patty, and gravy for the English muffin, Canadian bacon, and Hollandaise sauce.

Irish Benedict: Replace the Canadian bacon with corned beef hash.

Eggs Neptune: Substitute crab meat for the Canadian bacon.

33

 If you use quinoa as feed, rinse it first to remove the sapogenin—this tastes bad to birds.

34 Clamp the roof and let dry. The adhesive needed a couple of days to set up, so I clamped the edges down with a caul and a simple spring clamp. I secured the middle section of the roofing by nailing it directly to the sheathing at 1' (300mm) intervals.

35 Trim up the roofing. As with the felt paper (U), it is much easier to roll out the roofing material (T) if you have some extra width—the excess can be trimmed off with a utility knife.

36 Cover the windows. To seal up the windows, I cut a pair of square panels (AA) out of ¾" (17mm) plywood and attached them to the frames (J) with hinges. A little decoration never hurts, either—I used a scrap of wood and a chicken stamp I had.

37 Examine the window covering. The hinges on the left side of the window cover (AA) look neat and clean.

38 Shape the handles. As I wound up the process, I turned my attention to designing a set of handles that would enable the coop to be moved around easily. I used a pair of 40" (1015mm)-long 2x2s and rounded over their edges with a router to make them easy on the hands.

39 Attach the handles. The handles (S) are easily attached to the coop by drilling a pair of 1¾" (44mm) holes in the bottom corners of the end wall (D). The handles are then inserted and screwed into the floor of the coop with a couple of long 2" (50mm) screws.

40 Examine the handles. The limiting factor seemed to be the distance between the handles—if they were closer together, one person would have an even easier time maneuvering it, but that wasn't really much of an option because the door takes up most of the real estate in that area. Even so, it worked quite well.

Over hard = A fried egg with no runniness.

KATE & MOLLY
Salt Lake City, Utah

Kate and Molly have been raising chickens in their urban backyard for a little over a year. They began thinking about it during what Kate referred to as the first wave of enthusiasm for backyard chickens in their area: they knew a few people who were already doing it, but the huge rise in popularity in their area hadn't quite hit yet. Kate was pretty excited from the get-go, but Molly was a bit more hesitant: it seemed like a big leap, but over time it began to feel like a natural next step. After all, they were already very experienced gardeners who had been putting a lot of time into shaping up their yard, and the idea of producing more of their own food wasn't too much of a stretch. Molly was finally sold after she tasted fresh eggs for the first time.

history

This is Kate's first experience owning chickens as an adult, but she has a lot of great memories of them as a child. Her aunt raised chickens, and Kate remembers going into the coop when she was young and playing with the chickens. Agriculture is a long-standing tradition in her family—she recounts that her grandparents had a farm with sheep, corn, and orchards, and they really lived off the land. Her great-

⬆ Henny Penny is getting down to business in the nest box. Kate and Molly get about four eggs a day from their five hens.

⬅ The wire "roof" of the chicken run is covered with ivy. This provides some nice shade during the sunny summer months. When fall comes, Kate and Molly put corrugated galvanized tin over the wire to keep snow from piling up inside the run. In the meantime, though, the chickens love to hop up onto the top edge of the door to nibble at the ivy hanging within reach.

↑ Molly and Kate list fresh eggs and improved compost as the top two benefits of raising their own chickens. During the chillier winter months, they put a tarp over the coop enclosure to help keep snow and ice out— it's a simple solution to what otherwise becomes a very muddy situation.

For a special breakfast, fry up some bacon and then cook the egg in the bacon grease.

grandfather had a large chicken farm with thousands of birds during the Great Depression, and he used to supplement their income by selling chickens. So, having chickens now is a neat way of reconnecting to her family's history.

starting out

When the pair decided to get their own chickens about a year and a half ago, they bought day-old chicks from a local farm store. Molly remembers thinking how helpless they looked, and wondering if they could keep them alive. Although it took some time, Kate and Molly became confident they could meet the chicks' needs.

challenges

One early challenge arose when they needed to go out of town: the chicks were housed in a large cardboard box at that time, and the box and all of the chicks' supplies had to be hauled to Kate's brother's house so they could be cared for. That was a pain, but it worked out fine.

Another challenge was introducing the growing chicks to their new home outdoors when they outgrew their cardboard box. Molly recalls that they did this gradually, by bringing the chicks out for short periods and eventually working up to living outdoors full-time.

In addition to their flock of chickens, Kate and Molly have two dogs, and getting them all integrated took a little time. They soon discovered that one

of the dogs was scared of the chickens, which makes for a benign situation, whereas the other is prone to chasing them, so she needs close supervision when the chickens are wandering around outside of the run.

As is often the case, one of their day-old chicks turned out to be a rooster. Kate called around and found that a local community-supported working farm would be happy to take him in. Kate found a replacement hen on Craigslist,

⬆ Kate and Molly swear by this double-walled metal watering station—they find that it keeps water cooler in the summer months, and its large size means they don't have to fill it very often.

and after an adjustment period of a week or so, the new bird settled in just fine.

Having chickens has been a real learning opportunity as Kate and Molly find out what works for them and what doesn't. For one thing, they plan to put a solid roof over the run before this winter so that they can keep the snow out—they found that the chicken run can get pretty mucky when the snow melts. In an effort to keep it clean and dry, they put straw down, but they quickly saw that didn't help at all—it turned out that straw actually holds the moisture in. I've experienced the same frustration, and I suggested that they use sawdust instead, which works like a charm.

benefits

Kate and Molly cite lots of benefits to having chickens, in addition to the bounty of fresh eggs. Molly was surprised at how much fun it is just to watch them go about their daily routines—who knew

they would be so cute! And because they handled their chicks from day one, the adult chickens are used to it. This means that kids are able to come over and pick them up and have a lot of fun, which is pretty rewarding for everybody. Both Kate and Molly have enjoyed meeting other chicken owners and bonding over the experience, and they were happy to see the local ordinances changed to favor more backyard flocks. As Kate remarked at the end of our interview, "I don't know why everybody doesn't do it!"

tips

Here are some of Kate and Molly's tips for new chicken owners:

- **Watering stations.** Use a double-walled metal watering station; it keeps water cooler in the summer and seems to slow or even prevent the growth of algae in the water. Get the biggest one you can!

↙ Sometimes the little things make all the difference: putting a spring on the backside of the door ensures that it closes automatically, and this keeps the chickens from darting out, which they will do if given the smallest opportunity.

↓ These tall galvanized trash cans are perfect for storing food and bedding materials.

 Whether the eggs are white, brown, or blue, the nutritional value is the same.

- **Chicken poop.** Chicken poop is actually not a big deal, despite Kate's early worries that there would be a lot of it, and that it would be smelly. It breaks down quickly and a small home flock doesn't really produce that much of it.

- **Gardening.** Chickens and gardens go together perfectly: the birds will happily eat anything that people don't want, and they help to create compost to feed the garden's next growing cycle. Kate lets their chickens roam around on the compost pile, and says that they do a remarkably good job of turning it over and mixing up the contents.

⬆ The black and gold chickens are Sex Links, the grey and white is a Barred Rock, and the other two brown ones are Easter Eggers—they lay colored eggs like Ameraucanas do, but aren't pure-bred.

⬅ The coop, run, and garden are in close proximity—that way it is easy to move the manure to the compost heap, and the compost to the garden beds.

- **Wintering.** Unless you live in an extremely cold place, winters are easier to deal with than expected. Situate your coop in a sunny spot, and let the birds come out every day to peck around. Heat lamps aren't necessary in Kate and Molly's area.

- **Sizing the run.** Make sure you have a large enough run for the chickens to be able to hang out in without needing to let them out into the backyard—that takes a lot of time and supervision, especially if you have other pets that might pose a threat.

- **When the chickens stop laying.** They've have tossed around the question of what to do when the chickens become old and stop laying—the options are to eat them or keep them as non-productive pets—but the birds are only in their second year now, so they'll be able to cross that bridge when they come to it.

⬆ As an avid gardener, Kate often has plenty of extra produce to share with the birds, and they always rally around for a special treat.

 Rhode Island Reds are popular brown egg-layers and are very resilient.

COOP RUNS

Unless you have the area where you want to restrict your chickens completely fenced in already, you will need to build a chicken run around your coop. You will need some sturdy wood, preferably 2x4s, chicken wire or netting, and screws. There are many ways to build a run, so I will show you a few of the methods you can use. Keep in mind that utilizing existing fences and barriers will make your run a lot more stable and easier to build—be sure to take this into account when planning your design.

LOCATING A CHICKEN RUN ALONG A FENCE

Anytime I'm advising newbie chicken owners about where to locate their coop, I generally have a bit of a bias toward placing it against a fence. This is for economy, simplicity, and stability: rather than having to fence in all four sides of a run, it is easier and cheaper to just fence in two or three sides. This results in a notable savings of both time and money, but it also has the advantage of providing a really strong run (assuming that the fence was stable to begin with!). I say this because I've constructed a number of freestanding, four-sided chicken runs, and even when I've used galvanized framing supports, it is harder than you might think to build a rock-solid structure. So, when this option works for people, I do like to integrate a fence into a run. There is another factor to consider, however, and that is the fact that some local ordinances prohibit situating a chicken coop or run immediately adjacent to a property line. I know that some municipalities require a setback of 4' (1200mm), so it always pays to check out the requirements in your area.

1 **Frame out the panels.** Using 2x4s or 2x2s, construct rectangular frames. It's a good idea to make the walls 5' (1500mm) tall if you're not planning to add a top panel. It's also wise to break up any long distances into multiple frames for strength.

2 **Attach corner reinforcement plates.** The walls of the run can be fastened together in a couple of different ways—I like to make large plates like this with my band saw. They are a great way to use up scrap wood and they allow me to make really strong connections wherever two or more pieces of lumber meet. Use 3" (75mm) all-weather screws to attach the pieces.

3 **Add chicken wire.** Use a staple gun to attach chicken wire on all the panels.

Measuring lengths of chicken wire can be harder than it might seem because it has a tendency to want to curl back into the shape of the roll it came in. Putting a rock on one end and then stepping on it to flatten it helps.

Tin snips are the best tool I've found for cutting wire mesh.

Sometimes a panel is wider than the roll of chicken wire, in which case you'll have to overlap two layers of wire. I like to use plastic zip-ties every 18" (458mm) or so to seal up this overlap.

 Broody: A chicken who wants to sit on her eggs.

4 **Put the walls in place.** When I work by myself, I use a couple of 2x4s as temporary diagonal braces to keep the walls up.

5 **Join the walls together.** If you are planning a long wall with a door in the middle, like this one, joining the sections with a long 2x4 that spans the door opening is a good tactic.

6 **Attach the wall to the fence.** If there is an existing fence making up part of the run, use brackets to attach the new panels to the existing fence.

7 **Anchor the run.** Even if you are building a freestanding run, you should at least anchor the run to the fence for long-term stability.

8 Make and install the run door. In the same manner as you built the wall panels, make the door to fit the gap left for it. Use hardy hinges to attach the door to the run, and don't forget to install a latch of some kind.

You can also recycle old windows or doors to use as the run door.

9 Stabilize long panels. You will need to reinforce weak areas, most likely where you are spanning a great distance or next to a door opening. Cut a pair of angled braces and install them to strengthen the wall immensely.

10 If desired, add roof. Installing wire mesh on top of the roof is a lot easier with two people, but if you have to do it alone, it's not impossible. I started by stapling one end into place and then stretching it as tightly as possible across the span. You'll wish you had three hands, but once you get a few staples into the other end, it'll go ok. You could also install corrugated metal or plastic over the top of the run for shade.

8

9

10

Bibimbap, Korean: Warm rice; sautéed tofu, cucumber, and mushrooms; lettuce; chili paste; and a fried egg.

11

11 Fence in the bottom, if desired.
If you have a coop that exits from the bottom, you could opt to fence around the legs and create a short run to the side. If you want to do this, staple the wire to one leg and wrap it the whole way around. Instructions on building a short run are on page 141.

12 Fence around the nest box opening, if desired. Sometimes, you will want to access the nest box sidecar from outside the run. If this is what you want, you'll have to situate the coop adjacent to a wall panel. Trim the wire carefully around the box and staple it to the end panel of the coop.

12

CHICKEN TRACTOR

A chicken tractor is a short fenced-in box that allows backyard farmers to confine their chickens to a specific area that can change each day. In this way, you can control which areas of your land are fertilized and provide your chickens with fresh foraging plots each day. You can also use this as a short run (for example,

this works very well with Coop #2, page 28). Use 2x4 pieces or plywood plates and all-weather screws to attach the assembly to your coop securely. Make sure to install extra chicken wire as needed to close any gaps between the run and the coop.

1 **Build the run frames.** This design is quick and easy to make. I used 2x2s to make a rectangular frame for the top, and then I built a matching one for the bottom. The frames are 5' x 3' (1500 x 900mm).

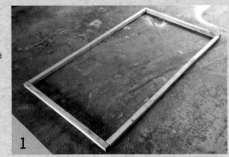

2 **Attach scrap plywood overlays.** The corner joints are really strong when the parts are assembled with simple plywood overlays. I used ½" (11mm) plywood, although you could use thicker stuff if you happen to have some on hand.

3 **Cover the top with chicken wire.** I covered the top with chicken wire. It is a bit of a pain to roll it out and staple it down when you're working by yourself, but it isn't too bad. Stapling down one end helps to keep it from rolling back up on itself as you go.

4 **Add frame legs.** I attached a set of four 16" (405mm)-tall legs to the bottom frame by screwing through the corner blocks. The result was a surprisingly strong frame.

5 **Attach the top and bottom.** I flipped the bottom frame upside down so that I could easily attach the top. And voila, you can see how quickly this goes together. I had originally planned to reinforce the corners even more, but it just didn't end up being necessary. This assembly is really robust. Staple chicken wire to the sides and your tractor is ready to use.

Be sure to provide enough grit for the chickens—it helps to grind up their food.

egging you on:
MORE COOP IDEAS

I dug into my photo archives to find more coop examples to get your brain humming. Scratch through these pages to unveil even more inspiration for creating and decorating your chicken coop! Be creative; think of a theme and carry it through all the details. If you like the ocean, perhaps you could design your coop to look like a beach house on stilts, with a surfboard-shaped ladder and cut-out wave silhouettes around the base. What about a Noah's ark–themed coop, or a coop that looks just like your house in miniature? Make the coop your own—you and your chickens are going to be spending a lot of time around it, so you might as well do it up right!

Coop Extras

❑ **Windows**
   ❑ Frames, joined with butt or miter joints
   ❑ Panes, made with wood pieces laid over the screen
   ❑ Awnings, made from wood or stretched canvas over a frame
   ❑ Shutters, flat or with slats, decorative or functional
   ❑ Vent cover, made with slats
   ❑ Glass or Plexiglas insert for cold weather

❑ **Doors**
   ❑ Raised panel look, made with router detailing
   ❑ Decorative knobs, upcycled from old cabinets
   ❑ Faux decorative hinges, painted on
   ❑ Doorknocker, salvaged or painted on
   ❑ Fanlights, cut out or painted on

❑ **Roof**
   ❑ Cupola, decorative or functional
   ❑ Weather vane, perhaps with a rooster

❑ **Interior**
   ❑ Poop trays, as on page 125
   ❑ Multiple levels, as on page 28
   ❑ Natural roosts, made from branches

❑ **Exterior details**
   ❑ Fish scale shingles
   ❑ Wood cut-outs: scallops, Victorian beadwork, sun rays, stars, etc.
   ❑ Colorful paint scheme
   ❑ Painted or stamped accents

⊕ This colorful and rustic coop features a multicolored sun motif, several stars, weathered boards on the doors, and a deluxe extra-wide chicken ladder fashioned from plywood scraps. The sturdy 2x2 legs raise the coop and provide ample shade and hiding spots underneath.

Chalazae: In an egg, two white cords that keep the yolk centered in the egg white.

⬆ Coops are fun to build because just about anything goes—you can keep it basic and utilitarian, or you can opt to dress up the design a little bit with paint and trim. These traditional motifs—a sunburst, a star, and a pair of half-moons—are all this simple coop needed.

● This layout will work great for a small flock of two or three birds. Both the roosts and the nest box are removable for easy cleanout. And, knowing what I do about chickens, I predict that the birds will also like to roost on the top of the nest box, given their penchant for hanging out in high places.

● By providing a small closeable access door on the side of the coop (which will be paired with a chicken ladder), the main doors on the front of the coop can be left closed most of the time and the occupants will be able to come and go at will and still have a relatively private place to come home to.

 Add crushed oyster shell to the feed to add calcium and improve shell thickness.

↑ I found these doors at a local resale shop for $4 each, and then cut out the half moon shapes and added a coat of paint. Pretty simple! The hinges and latches came from the dollar store, and will do the trick just fine. This latch could be further secured with a carabiner to ensure that even the cleverest of raccoons can't get in.

⬆ This coop features a nest box sidecar that is accessible from the back side of the coop. The lid also has a mechanical stay on it that locks into place when the lid is lifted so you can have both hands free. For $5, it's a neat option.

🥄 Oeufs a la Neige, French: A dessert with lightly poached meringue balls floating in vanilla custard.

⬆ This coop owner created some extra shade by draping a heavy cloth over part of the run. This kind of approach is neat because it is easily reversible in the cold winter months when direct sun would actually be a benefit. Also highlighted here are nice door pulls and lovely scalloped details on the coop itself.

⬆ Extra features, such as windows, aren't really necessary most of the time, but they provide a fun way to dress up the coop's exterior. If you have some scrap glass on hand, it won't really add to the cost, either.

 Put a few golf balls in the nest box to encourage your chickens to lay there.

⬆ This self-contained setup is located in a shady cluster of fruit trees—you can bet that the chickens will love chowing down on any fruit that hits the ground. Notice the unique roof profile, scrolled plywood on the legs, and molding around the doors.

 This run didn't require a lot of materials because it was situated at the back of the yard where the fence turns a corner. This kept the budget down and saved time, and the fences provide a lot of shade, which is great in the summer. Notice the corrugated metal on the roof.

Sprinkle food-grade diatomaceous earth in the run and coop to kill mites and lice.

⬆ Is it a playhouse or a chicken coop? This guy couldn't wait to check it out when I delivered it, but once the chickens made it home, I'm guessing he was ready to find someplace else to play. This sturdy coop features a drawbridge-style ladder and some nice decorative details.

⬆ The sunlight gleams off of the metal roof on this coop. I salvaged this material from a building site and kept it out of a landfill, and it will provide many years of service in its new role. You can use a lot of different things to roof a coop—as long as it is watertight and durable, you're good to go. Also notice the scrolled decorative panel on the legs and the matching blue molding around the window.

Encourage dust bathing by providing a small area of loose, dusty ground.

⬆ If you keep other animals as well, consider building your coop so there is ample space underneath—this area provides great shade during the summer months.

⬅ Nest boxes don't have to be fancy. In fact, they almost never are—scrap wood and butt joints fastened with screws or nails usually do the trick. Chickens seem to want a secure, semi-private place to lay eggs, and that is their main requirement. This nest box has a small raised lip on the front side so that chickens are less likely to push the bedding material out.

 If you want to get even more eco-friendly, consider building a rainwater catcher to store rainwater funneled from your roof. You can install a tap on the bottom for easy water dispensing. Recycling water like this will save you money and help the environment—and stored rainwater is great for watering plants and washing down outdoor stuff.

Leghorns are popular white egg-layers, though they have nervous personalities.

INDEX

ACQUISITION EDITOR
Peg Couch

COPY EDITOR
Paul Hambke

COVER AND LAYOUT DESIGNER
Lindsay Hess

EDITOR
Kerri Landis

EDITORIAL COORDINATOR
Heather Stauffer

PHOTOGRAPHY EDITOR
Scott Kriner

PROOFREADER
Lynda Jo Runkle

INDEXER
Jay Kreider

Loco Moco, Hawaiian: Rice topped with a hamburger patty, fried egg, and gravy.

Tree Craft
35 Rustic Wood Projects That Bring the Outdoors In
By Chris Lubkemann

Beautify your home with rustic accents made from twigs and branches. More than 35 eco-chic projects for a coat rack, curtain rods, candle holders, desk sets, picture frames, a table, chess set, and more.

ISBN: 978-1-56523-455-0
$19.95 · 128 Pages

Little Book of Whittling
Passing Time on the Trail, on the Porch, and Under the Stars
By Chris Lubkemann

Unwind while you learn to create useful and whimsical objects with nothing more than a pocket knife, a twig, and a few minutes of time.

ISBN: 978-1-56523-274-7
$12.95 · 104 Pages

Handmade Music Factory
The Ultimate Do-It-Yourself Guide to Foot-Stompin Good Instruments
By Mike Orr

Make instruments that look great—and sound better—from a one-string soup can guitar to a hubcap banjo, and even a stand-up guitar made from a vintage ironing board.

ISBN: 978-1-56523-559-5
$22.95 · 160 Pages

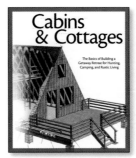

Cabins and Cottages
The Basics of Building a Getaway Retreat for Hunting, Camping, and Rustic Living
By Skills Institute Press

Building a rustic cottage or log cabin is a project any DIYer or woodworker would like to tackle. This book gives you the complete know-how, from clearing a site to raising the walls and adding the amenities.

ISBN: 978-1-56523-539-7
$19.95 · 160 Pages

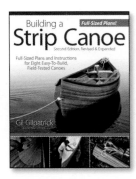

Building a Strip Canoe, Second Edition, Revised & Expanded
Full-Sized Plans and Instructions for Eight Easy-To-Build, Field-Tested Canoes
By Gil Gilpatrick

Paddle along with an expert outdoorsman and canoe builder as he shares his experience in guiding both novice and accomplished woodworkers in building a canoe with easy step-by-step instructions.

ISBN: 978-1-56523-483-3
$24.95 · 112 Pages

Homeowner's Guide to the Chainsaw
A chainsaw pro shows you how to safely and confidently handle everything from trimming branches and felling trees to splitting and stacking wood
By Brian and Jen Ruth

Written by professional chainsaw carvers, this resource teaches homeowners how to safely complete everyday yard work, such as grooming hedges, cutting firewood, and trimming tree limbs, with a chainsaw.

ISBN: 978-1-56523-356-0
$24.95 · 200 Pages

Get to know a variety of farm animals with this series of primer guides to breeds

Know Your Chickens
By Jack Byard

Forty-four breeds of chicken—some rare, each with a rich diversity of color, size and feather pattern.

ISBN: 978-1-56523-612-7
$8.95 · 96 Pages

Know Your Cattle
By Jack Byard

Forty-four breeds of cattle—many you will recognize, but some are very rare.

ISBN:978-1-56523-613-4
$8.95 · 96 Pages

Know Your Pigs
By Jack Byard

Twenty-eight breeds of pigs—from the American Guinea Hog to the Wild Boar.

ISBN: 978-1-56523-611-0
$8.95 · 64 Pages

Know Your Donkeys
By Jack Byard

Thirty-five breeds of donkey—from the miniature to the mammoth with a couple of mules thrown in.

ISBN: 978-1-56523-614-1
$8.95 · 80 Pages

More great woodworking books from Fox Chapel Publishing

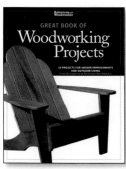

Great Book of Woodworking Projects
50 Projects for Indoor Improvements and Outdoor Living from the experts at American Woodworker
Edited by Randy Johnson

An ideal resource for woodworkers looking for a new project or wanting to spruce up their home, with plans for 50 projects that can take a few hours or up to a weekend to complete.

ISBN: 978-1-56523-504-5
$24.95 · 256 Pages

Outdoor Furniture
14 Timeless Woodworking Projects for the Yard, Deck, and Patio
By Skills Institute Press

Design and build beautiful wooden outdoor furniture so you can admire your new chicken coop in style.

ISBN: 978-1-56523-500-7
$19.95 · 144 Pages

How to Make Bookshelves & Bookcases
19 Outstanding Storage Projects from the Experts at American Woodworker
Edited by Randy Johnson

Build functional yet stylish pieces from a simple wall shelf to a grand bookcase. The experts at American Woodworker give step-by-step instructions using various types of wood.

ISBN: 978-1-56523-458-1
$19.95 · 184 Pages

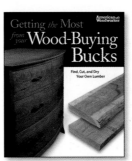

Getting the Most from your Wood-Buying Bucks
Find, Cut, and Dry Your Own Lumber
Edited by Tom Caspar

An essential guide for every woodworker on how to purchase wood wisely for a home workshop. This skill-expanding, money-saving book includes expert advice, detailed drawings, and step-by-step photographs.

ISBN: 978-1-56523-460-4
$19.95 · 208 Pages

Big Book of Brewing
The Classic Guide to All-Grain Brewing
By Dave Line

Interested in home brewing but don't know where to start? The easy to understand instructions from this master brewer makes the process simple and easy.

ISBN: 978-1-56523-603-5
$17.95 · 256 Pages

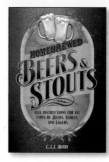

Homebrewed Beers & Stouts
Full Instructions for All Types of Beers, Stouts, and Lagers
By C.J.J. Berry

Learn how to brew your own great-tasting beer at home with more than 70 recipes, from a light summer ale to an authentic stout.

ISBN: 978-1-56523-601-1
$14.95 · 160 Pages

Real Cidermaking on a Small Scale
An Introduction to Producing Cider at Home
By Michael Pooley and John Lomax

Learn everything you need to know about the process of making hard cider in your home from any kind of apple.

ISBN: 978-1-56523-604-2
$12.95 · 136 Pages

130 New Winemaking Recipes
How to Succeed with Home-Made Wines from Fruits, Grains, and Herbs
By C.J.J. Berry

Filled with 130 recipes utilizing traditional country ingredients, this book is a must-have for anyone who has discovered the rewards of at-home winemaking.

ISBN: 978-1-56523-600-4
$12.95 · 128 Pages

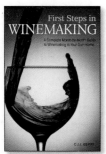

First Steps in Winemaking
A Complete Month-by-Month Guide to Winemaking in Your Own Home
By C.J.J. Berry

Become an at-home vintner using the methods and techniques in this book. Learn which wines are best for which seasons.

ISBN: 978-1-56523-602-8
$14.95 · 240 Pages

Treasured Amish and Mennonite Recipes
A Taste of the Simple Life
By Mennonite Central Committee

Bring these authentic Amish and Mennonite recipes into your kitchen—includes classics such as chicken pot-pie and mashed potatoes.

ISBN: 978-1-56523-599-1
$19.95 · 320 Pages